Introduction to
Stochastic Calculus
Applied to Finance

Introduction to
Stochastic Calculus
Applied to Finance

Damien Lamberton
L'Université de Marne la Vallée
France

and

Bernard Lapeyre
L'Ecole Nationale des Ponts et Chaussées
France

Translated by

Nicolas Rabeau
Centre for Quantitative Finance
Imperial College, London
and
Merrill Lynch Int. Ltd., London

and

François Mantion
Centre for Quantitative Finance
Imperial College
London

CHAPMAN & HALL/CRC

Boca Raton   London   New York   Washington, D.C.

Library of Congress Cataloging-in-Publication Data

Catalog record is available from the Library of Congress.

First edition 1996
First CRC Press reprint 2000

© 1996 by Chapman & Hall/CRC

No claim to original U.S. Government works
International Standard Book Number 0-412-71800-6
Printed in the United States of America   1   2   3   4   5   6   7   8   9   0
Printed on acid-free paper

Contents

Introduction

The objective of this book is to give an introduction to the probabilistic techniques required to understand the most widely used financial models. In the last few years, financial quantitative analysts have used more sophisticated mathematical concepts, such as martingales or stochastic integration, in order to describe the behaviour of markets or to derive computing methods.

In fact, the appearance of probability theory in financial modelling is not recent. At the beginning of this century, Bachelier (1900), in trying to build up a 'Theory of Speculation', discovered what is now called *Brownian motion*. From 1973, the publications by Black and Scholes (1973) and Merton (1973) on option pricing and hedging gave a new dimension to the use of probability theory in finance. Since then, as the option markets have evolved, Black-Scholes and Merton results have developed to become clearer, more general and mathematically more rigorous. The theory seems to be advanced enough to attempt to make it accessible to students.

Options

Our presentation concentrates on options, because they have been the main motivation in the construction of the theory and still are the most spectacular example of the relevance of applying stochastic calculus to finance. An option gives its holder the right, but *not the obligation*, to buy or sell a certain amount of a financial asset, by a certain date, for a certain strike price.

The writer of the option needs to specify:

- the type of option: the option to buy is called a *call* while the option to sell is a *put*;
- the underlying asset: typically, it can be a stock, a bond, a currency and so on;

- the amount of an underlying asset to be purchased or sold;

- the expiration date: if the option can be exercised at any time before maturity, it is called an *American* option but, if it can only be exercised at maturity, it is called a *European* option;

- the exercise price which is the price at which the transaction is done if the option is exercised.

The price of the option is the *premium*. When the option is traded on an organised market, the premium is quoted by the market. Otherwise, the problem is to price the option. Also, even if the option is traded on an organised market, it can be interesting to detect some possible abnormalities in the market.

Let us examine the case of a European call option on a stock, whose price at time t is denoted by S_t. Let us call T the expiration date and K the exercise price. Obviously, if K is greater than S_T, the holder of the option has no interest whatsoever in exercising the option. But, if $S_T > K$, the holder makes a profit of $S_T - K$ by exercising the option, i.e. buying the stock for K and selling it back on the market at S_T. Therefore, the value of the call at maturity is given by

$$(S_T - K)_+ = \max(S_T - K, 0).$$

If the option is exercised, the writer must be able to deliver a stock at price K. It means that he or she must generate an amount $(S_T - K)_+$ at maturity. At the time of writing the option, which will be considered as the origin of time, S_T is unknown and therefore two questions have to be asked:

1. How much should the buyer pay for the option? In other words, how should we price at time $t = 0$ an asset worth $(S_T - K)_+$ at time T? That is the problem of *pricing* the option.

2. How should the writer, who earns the premium initially, generate an amount $(S_T - K)_+$ at time T? That is the problem of *hedging* the option.

Arbitrage and put/call parity

We can only answer the two previous questions if we make a few necessary assumptions. The basic one, which is commonly accepted in every model, is the absence of arbitrage opportunity in liquid financial markets, i.e. there is no riskless profit available in the market. We will translate that into mathematical terms in the first chapter. At this point, we will only show how we can derive formulae relating European put and call prices. Both the put and the call which have maturity T and exercise price K are contingent on the same underlying asset which is worth S_t at time t. We shall assume that it is possible to borrow or invest money at a constant rate r.

Let us denote by C_t and P_t respectively the prices of the call and the put at time t. Because of the absence of arbitrage opportunity, the following equation called *put/call parity* is true for all $t < T$

$$C_t - P_t = S_t - Ke^{-r(T-t)}.$$

To understand the notion of arbitrage, let us show how we could make a riskless profit if, for instance,

$$C_t - P_t > S_t - Ke^{-r(T-t)}.$$

At time t, we purchase a share of stock and a put, and sell a call. The net value of the operation is

$$C_t - P_t - S_t.$$

If this amount is positive, we invest it at rate r until time T, whereas if it is negative we borrow it at the same rate. At time T, two outcomes are possible:

- $S_T > K$: the call is exercised, we deliver the stock, receive the amount K and clear the cash account to end up with a wealth $K + e^{r(T-t)}(C_t - P_t - S_t) > 0$.

- $S_T \leq K$: we exercise the put and clear our bank account as before to finish with the wealth $K + e^{r(T-t)}(C_t - P_t - S_t) > 0$.

In both cases, we locked in a positive profit without making any initial endowment: this is an example of an arbitrage strategy.

There are many similar examples in the book by Cox and Rubinstein (1985). We will not review all these formulae, but we shall characterise mathematically the notion of a *financial market without arbitrage opportunity*.

Black-Scholes model and its extensions

Even though no-arbitrage arguments lead to many interesting equations, they are not sufficient in themselves for deriving pricing formulae. To achieve this, we need to model stock prices more precisely. Black and Scholes were the first to suggest a model whereby we can derive an explicit price for a European call on a stock that pays no dividend. According to their model, the writer of the option can hedge himself perfectly, and actually the call premium is the amount of money needed at time 0 to replicate exactly the payoff $(S_T - K)_+$ by following their dynamic hedging strategy until maturity. Moreover, the formula depends on only one non-directly observable parameter, the so-called *volatility*.

It is by expressing the profit and loss resulting from a certain trading strategy as a stochastic integral that we can use stochastic calculus and, particularly, Itô formula, to obtain closed form results. In the last few years, many extensions of the Black-Scholes methods have been considered. From a thorough study of the Black-Scholes model, we will attempt to give to the reader the means to understand those extensions.

Contents of the book

The first two chapters are devoted to the study of discrete time models. The link between the mathematical concept of martingale and the economic notion of arbitrage is brought to light. Also, the definition of complete markets and the pricing of options in these markets are given. We have decided to adopt the formalism of Harrison and Pliska (1981) and most of their results are stated in the first chapter, taking the Cox, Ross and Rubinstein model as an example.

The second chapter deals with American options. Thanks to the theory of optimal stopping in a discrete time set-up, which uses quite elementary methods, we introduce the reader to all the ideas that will be developed in continuous time in subsequent chapters.

Chapter 3 is an introduction to the main results in stochastic calculus that we will use in Chapter 4 to study the Black-Scholes model. As far as European options are concerned, this model leads to explicit formulae. But, in order to analyse American options or to perform computations within more sophisticated models, we need numerical methods based on the connection between option pricing and partial differential equations. These questions are addressed in Chapter 5.

Chapter 6 is a relatively quick introduction to the main interest rate models and Chapter 7 looks at the problems of option pricing and hedging when the price of the underlying asset follows a simple jump process.

In these latter cases, perfect hedging is no longer possible and we must define a criterion to achieve optimal hedging. These models are rather less optimistic than the Black-Scholes model and seem to be closer to reality. However, their mathematical treatment is still a matter of research, in the framework of so-called *incomplete markets*.

Finally, in order to help the student to gain a practical understanding, we have included a chapter dealing with the simulation of financial models and the use of computers in the pricing and hedging of options. Also, a few exercises and longer questions are listed at the end of each chapter.

This book is only an introduction to a field that has already benefited from considerable research. Bibliographical notes are given in some chapters to help the reader to find complementary information. We would also like to warn the reader that some important questions in financial mathematics are not tackled. Amongst them are the problems of optimisation and the questions of equilibrium for which the reader might like to consult the book by D. Duffie (1988).

A good level in probability theory is assumed to read this book. The reader is referred to Dudley (1989) and Williams (1991) for prerequisites. However, some basic results are also proved in the Appendix.

Acknowledgements

This book is based on the lecture notes taught at *l'Ecole Nationale des Ponts et Chaussées* since 1988. The organisation of this lecture series would not have been possible without the encouragement of N. Bouleau. Thanks to his dynamism, CERMA (Applied Mathematics Institute of ENPC) started working on financial modelling as early as 1987, sponsored by *Banque Indosuez* and subsequently by *Banque Internationale de Placement*.

Since then, we have benefited from many stimulating discussions with G. Pages and other academics at CERMA, particularly O. Chateau and G. Caplain. A few people kindly read the earlier draft of our book and helped us with their remarks. Amongst them are S. Cohen, O. Faure, C. Philoche, M. Picqué and X. Zhang. Finally, we thank our colleagues at the university and at INRIA for their advice and their motivating comments: N. El Karoui, T. Jeulin, J.F. Le Gall and D. Talay.

1

Discrete-time models

The objective of this chapter is to present the main ideas related to option theory within the very simple mathematical framework of discrete-time models. Essentially, we are exposing the first part of the paper by Harrison and Pliska (1981). Cox, Ross and Rubinstein's model is detailed at the end of the chapter in the form of a problem with its solution.

1.1 Discrete-time formalism

1.1.1 Assets

A discrete-time financial model is built on a finite probability space $(\Omega, \mathcal{F}, \mathbf{P})$ equipped with a filtration, i.e. an increasing sequence of σ-algebras included in $\mathcal{F}$: $\mathcal{F}_0, \mathcal{F}_1, \ldots, \mathcal{F}_N$. $\mathcal{F}_n$ can be seen as the information available at time n and is sometimes called the *σ-algebra of events up to time n*. The *horizon N* will often correspond to the maturity of the options. From now on, we will assume that $\mathcal{F}_0 = \{\emptyset, \Omega\}$, $\mathcal{F}_N = \mathcal{F} = \mathcal{P}(\Omega)$ and $\forall \omega \in \Omega$, $\mathbf{P}(\{\omega\}) > 0$. The market consists in $(d + 1)$ financial assets, whose prices at time n are given by the non-negative random variables $S_n^0, S_n^1, \ldots, S_n^d$, measurable with respect to $\mathcal{F}_n$ (investors know past and present prices but obviously not the future ones). The vector $S_n = (S_n^0, S_n^1, \ldots, S_n^d)$ is the vector of prices at time n. The asset indexed by 0 is the *riskless asset* and we have $S_0^0 = 1$. If the return of the riskless asset over one period is constant and equal to r, we will obtain $S_n^0 = (1 + r)^n$. The coefficient $\beta_n = 1/S_n^0$ is interpreted as the discount factor (from time n to time 0): if an amount β_n is invested in the riskless asset at time 0, then one dollar will be available at time n. The assets indexed by $i = 1 \ldots d$ are called *risky assets*.

1.1.2 Strategies

A *trading strategy* is defined as a stochastic process (i.e. a sequence in the discrete case) $\phi = \left((\phi_n^0, \phi_n^1, \ldots, \phi_n^d) \right)_{0 \leq n \leq N}$ in $\mathbb{R}^{d+1}$ where ϕ_n^i denotes the number of shares of asset i held in the portfolio at time n. ϕ is predictable, i.e.

$$\forall i \in \{0, 1, \ldots, d\} \begin{cases} \phi_0^i \text{ is } \mathcal{F}_0\text{-measurable} \\ \\ \text{and, for } n \geq 1: \quad \phi_n^i \text{ is } \mathcal{F}_{n-1}\text{-measurable.} \end{cases}$$

This assumption means that the positions in the portfolio at time n $(\phi_n^0, \phi_n^1, \ldots, \phi_n^d)$ are decided with respect to the information available at time $(n-1)$ and kept until time n when new quotations are available.

The *value of the portfolio* at time n is the scalar product

$$V_n(\phi) = \phi_n . S_n = \sum_{i=0}^{d} \phi_n^i S_n^i.$$

Its *discounted value* is

$$\tilde{V}_n(\phi) = \beta_n (\phi_n . S_n) = \phi_n . \tilde{S}_n,$$

with $\beta_n = 1/S_n^0$ and $\tilde{S}_n = (1, \beta_n S_n^1, \ldots, \beta_n S_n^d)$ is the vector of discounted prices.

A strategy is called *self-financing* if the following equation is satisfied for all $n \in \{0, 1, \ldots, N-1\}$

$$\phi_n . S_n = \phi_{n+1} . S_n .$$

The interpretation is the following: at time n, once the new prices $S_n^0, \ldots, S_n^d$, are quoted, the investor readjusts his positions from ϕ_n to ϕ_{n+1} without bringing or consuming any wealth.

Remark 1.1.1 The equality $\phi_n . S_n = \phi_{n+1} . S_n$ is obviously equivalent to

$$\phi_{n+1} . (S_{n+1} - S_n) = \phi_{n+1} . S_{n+1} - \phi_n . S_n,$$

or to

$$V_{n+1}(\phi) - V_n(\phi) = \phi_{n+1} . (S_{n+1} - S_n).$$

At time $n + 1$, the portfolio is worth $\phi_{n+1} . S_{n+1}$ and $\phi_{n+1} . S_{n+1} - \phi_{n+1} . S_n$ is the net gain caused by the price changes between times n and $n + 1$. Hence, the profit or loss realised by following a self-financing strategy is only due to the price moves.

The following proposition makes this clear in terms of discounted prices.

Proposition 1.1.2 *The following are equivalent*

(i) *The strategy ϕ is self-financing.*

(ii) *For any $n \in \{1, \ldots, N\}$,*

$$V_n(\phi) = V_0(\phi) + \sum_{j=1}^{n} \phi_j \cdot \Delta S_j,$$

where ΔS_j is the vector $S_j - S_{j-1}$.

(iii) *For any* $n \in \{1, \ldots, N\}$,

$$\tilde{V}_n(\phi) = V_0(\phi) + \sum_{j=1}^{n} \phi_j \cdot \Delta \tilde{S}_j,$$

where $\Delta \tilde{S}_j$ *is the vector* $\tilde{S}_j - \tilde{S}_{j-1} = \beta_j S_j - \beta_{j-1} S_{j-1}$.

Proof. The equivalence between (i) and (ii) results from Remark 1.1.1. The equivalence between (i) and (iii) follows from the fact that $\phi_n.S_n = \phi_{n+1}.S_n$ if and only if $\phi_n.\tilde{S}_n = \phi_{n+1}.\tilde{S}_n$. □

This proposition shows that, if an investor follows a self-financing strategy, the discounted value of his portfolio, hence its value, is completely defined by the initial wealth and the strategy $(\phi_n^1, \ldots, \phi_n^d)_{0 \leq n \leq N}$ (this is only justified because $\Delta \tilde{S}_j^0 = 0$). More precisely, we can prove the following proposition.

Proposition 1.1.3 *For any predictable process* $((\phi_n^1, \ldots, \phi_n^d))_{0 \leq n \leq N}$ *and for any* $\mathcal{F}_0$-*measurable variable* V_0, *there exists a unique predictable process* $(\phi_n^0)_{0 \leq n \leq N}$ *such that the strategy* $\phi = (\phi^0, \phi^1, \ldots, \phi^d)$ *is self-financing and its initial value is* V_0.

Proof. The self-financing condition implies

$$\begin{aligned}
\tilde{V}_n(\phi) &= \phi_n^0 + \phi_n^1 \tilde{S}_n^1 + \cdots + \phi_n^d \tilde{S}_n^d \\
&= V_0 + \sum_{j=1}^{n} \left(\phi_j^1 \Delta \tilde{S}_j^1 + \cdots + \phi_j^d \Delta \tilde{S}_j^d \right)
\end{aligned}$$

which defines ϕ_n^0. We just have to check that ϕ^0 is predictable, but this is obvious if we consider the equation

$$\phi_n^0 = V_0 + \sum_{j=1}^{n-1} \left(\phi_j^1 \Delta \tilde{S}_j^1 + \cdots + \phi_j^d \Delta \tilde{S}_j^d \right) + \left(\phi_n^1 \left(-\tilde{S}_{n-1}^1 \right) + \cdots + \phi_n^d \left(-\tilde{S}_{n-1}^d \right) \right)$$

□

1.1.3 Admissible strategies and arbitrage

We did not make any assumption on the sign of the quantities ϕ_n^i. If $\phi_n^0 < 0$, we have borrowed the amount $|\phi_n^0|$ in the riskless asset. If $\phi_n^i < 0$ for $i \geq 1$, we say that we are *short* a number ϕ_n^i of asset i. Short-selling and borrowing is allowed but the value of our portfolio must be positive at all times.

Definition 1.1.4 *A strategy* ϕ *is admissible if it is self-financing and if* $V_n(\phi) \geq 0$ *for any* $n \in \{0, 1, \ldots, N\}$.

The investor must be able to pay back his debts (in riskless or risky asset) at any time.

The notion of *arbitrage* (possibility of riskless profit) can be formalised as follows:

Definition 1.1.5 *An arbitrage strategy is an admissible strategy with zero initial value and non-zero final value.*

Most models exclude any arbitrage opportunity and the objective of the next section is to characterise these models with the notion of martingale.

1.2 Martingales and arbitrage opportunities

In order to analyse the connections between martingales and arbitrage, we must first define a *martingale* on a finite probability space. The conditional expectation plays a central role in this definition and the reader can refer to the Appendix for a quick review of its properties.

1.2.1 Martingales and martingale transforms

In this section, we consider a finite probability space $(\Omega, \mathcal{F}, \mathbf{P})$, with $\mathcal{F} = \mathcal{P}(\Omega)$ and $\forall \omega \in \Omega$, $\mathbf{P}(\{\omega\}) > 0$, equipped with a filtration $(\mathcal{F}_n)_{0 \leq n \leq N}$ (without necessarily assuming that $\mathcal{F}_N = \mathcal{F}$, nor $\mathcal{F}_0 = \{\emptyset, \Omega\}$). A sequence $(X_n)_{0 \leq n \leq N}$ of random variables is adapted to the filtration if for any n, X_n is $\mathcal{F}_n$-measurable.

Definition 1.2.1 *An adapted sequence $(M_n)_{0 \leq n \leq N}$ of real random variables is:*

- *a martingale if $\mathbf{E}(M_{n+1}|\mathcal{F}_n) = M_n$ for all $n \leq N - 1$;*
- *a supermartingale if $\mathbf{E}(M_{n+1}|\mathcal{F}_n) \leq M_n$ for all $n \leq N - 1$;*
- *a submartingale if $\mathbf{E}(M_{n+1}|\mathcal{F}_n) \geq M_n$ for all $n \leq N - 1$.*

These definitions can be extended to the multidimensional case: for instance, a sequence $(M_n)_{0 \leq n \leq N}$ of $\mathbb{R}^d$-valued random variables is a martingale if each component is a real-valued martingale.

In a financial context, saying that the price $(S_n^i)_{0 \leq n \leq N}$ of the asset i is a martingale implies that, at each time n, the best estimate (in the least-square sense) of S_{n+1}^i is given by S_n^i.

The following properties are easily derived from the previous definition and stand as a good exercise to get used to the concept of conditional expectation.

1. $(M_n)_{0 \leq n \leq N}$ is a martingale if and only if

$$\mathbf{E}(M_{n+j}|\mathcal{F}_n) = M_n \quad \forall j \geq 0$$

2. If $(M_n)_{n \geq 0}$ is a martingale, thus for any n: $\mathbf{E}(M_n) = \mathbf{E}(M_0)$.

3. The sum of two martingales is a martingale.

4. Obviously, similar properties can be shown for supermartingales and submartingales.

Definition 1.2.2 *An adapted sequence* $(H_n)_{0 \leq n \leq N}$ *of random variables is predictable if, for all* $n \geq 1$, H_n *is* $\mathcal{F}_{n-1}$ *measurable.*

Proposition 1.2.3 *Let* $(M_n)_{0 \leq n \leq N}$ *be a martingale and* $(H_n)_{0 \leq n \leq N}$ *a predictable sequence with respect to the filtration* $(\mathcal{F}_n)_{0 \leq n \leq N}$. *Denote* $\Delta M_n = M_n - M_{n-1}$. *The sequence* $(X_n)_{0 \leq n \leq N}$ *defined by*

$$
\begin{aligned}
X_0 &= H_0 M_0 \\
X_n &= H_0 M_0 + H_1 \Delta M_1 + \cdots + H_n \Delta M_n \quad \text{for } n \geq 1
\end{aligned}
$$

is a martingale with respect to $(\mathcal{F}_n)_{0 \leq n \leq N}$.

(X_n) is sometimes called the *martingale transform* of (M_n) by (H_n). A consequence of this proposition and Proposition 1.1.2 is that if the discounted prices of the assets are martingales, the *expected value* of the wealth generated by following a self-financing strategy is equal to the initial wealth.

Proof. Clearly, (X_n) is an adapted sequence. Moreover, for $n \geq 0$

$$
\begin{aligned}
&\mathbf{E}\left(X_{n+1} - X_n | \mathcal{F}_n\right) \\
&= \mathbf{E}\left(H_{n+1}(M_{n+1} - M_n) | \mathcal{F}_n\right) \\
&= H_{n+1}\mathbf{E}\left(M_{n+1} - M_n | \mathcal{F}_n\right) \text{ since } H_{n+1} \text{ is } \mathcal{F}_n\text{-measurable} \\
&= 0.
\end{aligned}
$$

Hence

$$
\mathbf{E}\left(X_{n+1} | \mathcal{F}_n\right) = \mathbf{E}\{X_n | \mathcal{F}_n\} = X_n.
$$

That shows that (X_n) is a martingale. $\qquad\square$

The following proposition is a very useful characterisation of martingales.

Proposition 1.2.4 *An adapted sequence of real random variables* (M_n) *is a martingale if and only if for any predictable sequence* (H_n), *we have*

$$
\mathbf{E}\left(\sum_{n=1}^{N} H_n \Delta M_n\right) = 0.
$$

Proof. If (M_n) is a martingale, the sequence (X_n) defined by $X_0 = 0$ and, for $n \geq 1$, $X_n = \sum_{n=1}^{N} H_n \Delta M_n$ for any predictable process (H_n) is also a martingale, by Proposition 1.2.3. Hence, $\mathbf{E}(X_N) = \mathbf{E}(X_0) = 0$. Conversely, we notice that if $j \in \{1, \ldots, N\}$, we can associate the sequence (H_n) defined by $H_n = 0$ for $n \neq j + 1$ and $H_{j+1} = \mathbf{1}_A$, for any $\mathcal{F}_j$-measurable A. Clearly, (H_n) is predictable and $\mathbf{E}\left(\sum_{n=1}^{N} H_n \Delta M_n\right) = 0$ becomes

$$
\mathbf{E}\left(\mathbf{1}_A \left(M_{j+1} - M_j\right)\right) = 0.
$$

Therefore $\mathbf{E}\left(M_{j+1} | \mathcal{F}_j\right) = M_j$. $\qquad\square$

1.2.2 Viable financial markets

Let us get back to the discrete-time models introduced in the first section.

Definition 1.2.5 *The market is* viable *if there is no arbitrage opportunity.*

Lemma 1.2.6 *If the market is viable, any predictable process* $(\phi^1, \ldots, \phi^d)$ *satisfies*

$$\tilde{G}_N(\phi) \notin \Gamma.$$

Proof. Let us assume that $\tilde{G}_N(\phi) \in \Gamma$. First, if $\tilde{G}_n(\phi) \geq 0$ for all $n \in \{0, \ldots, N\}$ the market is obviously not viable. Second, if the $\tilde{G}_n(\phi)$ are not all non-negative, we define $n = \sup \left\{ k | \mathbf{P}\left(\tilde{G}_k(\phi) < 0 \right) > 0 \right\}$. It follows from the definition of n that

$$n \leq N - 1, \quad \mathbf{P}\left(\tilde{G}_n(\phi) < 0 \right) > 0 \text{ and } \forall m > n \quad \tilde{G}_m(\phi) \geq 0.$$

We can now introduce a new process ψ

$$\psi_j(\omega) = \begin{cases} 0 & \text{if } j \leq n \\ 1_A(\omega)\phi_j(\omega) & \text{if } j > n \end{cases}$$

where A is the event $\left\{ \tilde{G}_n(\phi) < 0 \right\}$. Because ϕ is predictable and A is $\mathcal{F}_n$-measurable, ψ is also predictable. Moreover

$$\tilde{G}_j(\psi) = \begin{cases} 0 & \text{if } j \leq n \\ 1_A \left(\tilde{G}_j(\phi) - \tilde{G}_n(\phi) \right) & \text{if } j > n \end{cases}$$

thus, $\tilde{G}_j(\psi) \geq 0$ for all $j \in \{0, \ldots, N\}$ and $\tilde{G}_N(\psi) > 0$ on A. That contradicts the assumption of market viability and completes the proof of the lemma. $\qquad \square$

Theorem 1.2.7 *The market is viable if and only if there exists a probability measure* $\mathbf{P}^*$ *equivalent* [†] *to* $\mathbf{P}$ *such that the discounted prices of assets are* $\mathbf{P}^*$-*martingales.*

Proof. (a) Let us assume that there exists a probability $\mathbf{P}^*$ equivalent to $\mathbf{P}$ under which discounted prices are martingales. Then, for any self-financing strategy (ϕ_n), (1.1.2) implies

$$\tilde{V}_n(\phi) = V_0(\phi) + \sum_{j=1}^{n} \phi_j . \Delta \tilde{S}_j.$$

Thus by Proposition 1.2.3, $\left(\tilde{V}_n(\phi) \right)$ is a $\mathbf{P}^*$- martingale. Therefore $\tilde{V}_N(\phi)$ and $V_0(\phi)$ have the same expectation under $\mathbf{P}^*$:

$$\mathbf{E}^* \left(\tilde{V}_N(\phi) \right) = \mathbf{E}^* \left(\tilde{V}_0(\phi) \right).$$

[†] Recall that two probability measures $\mathbf{P}_1$ and $\mathbf{P}_2$ are equivalent if and only if for any event A, $\mathbf{P}_1(A) = 0 \Leftrightarrow \mathbf{P}_2(A) = 0$. Here, $\mathbf{P}^*$ equivalent to $\mathbf{P}$ means that, for any $\omega \in \Omega$, $\mathbf{P}^*(\{\omega\}) > 0$.

If the strategy is admissible and its initial value is zero, then $\mathbf{E}^* \left(\tilde{V}_N(\phi) \right) = 0$, with $\tilde{V}_N(\phi) \geq 0$. Hence $\tilde{V}_N(\phi) = 0$ since $\mathbf{P}^*(\{\omega\}) > 0$, for all $\omega \in \Omega$.

(b) The proof of the converse implication is more tricky. Let us call Γ the convex cone of strictly positive random variables. The market is viable if and only if for any admissible strategy ϕ: $V_0(\phi) = 0 \Rightarrow \tilde{V}_N(\phi) \notin \Gamma$.

(b1) To any admissible process $(\phi_n^1, \ldots, \phi_n^d)$ we associate the process defined by

$$\tilde{G}_n(\phi) = \sum_{j=1}^n \left(\phi_j^1 \Delta \tilde{S}_j^1 + \cdots + \phi_j^d \Delta \tilde{S}_j^d \right).$$

That is the cumulative discounted gain realised by following the self-financing strategy $\phi_n^1, \ldots, \phi_n^d$. According to Proposition 1.1.3, there exists a (unique) process (ϕ_n^0) such that the strategy $((\phi_n^0, \phi_n^1, \ldots, \phi_n^d))$ is self-financing with zero initial value. $\tilde{G}_n(\phi)$ is the discounted value of this strategy at time n and because the market is viable, the fact that this value is positive at any time, i.e $\tilde{G}_n(\phi) \geq 0$ for $n = 1, \ldots, N$, implies that $\tilde{G}_N(\phi) = 0$. The following lemma shows that even if we do not assume that $\tilde{G}_n(\phi)$ are non-negative, we still have $\tilde{G}_N(\phi) \notin \Gamma$.

(b2) The set $\mathcal{V}$ of random variables $\tilde{G}_N(\phi)$, with ϕ predictable process in $\mathbb{R}^d$, is clearly a vector subspace of $\mathbb{R}^\Omega$ (where $\mathbb{R}^\Omega$ is the set of real random variables defined on Ω). According to Lemma 1.2.6, the subspace $\mathcal{V}$ does not intersect Γ. Therefore it does not intersect the convex compact set $K = \{X \in \Gamma | \sum_\omega X(\omega) = 1\}$ which is included in Γ. As a result of the convex sets separation theorem (see the Appendix), there exists $(\lambda(\omega))_{\omega \in \Omega}$ such that:

1. $\forall X \in K, \quad \sum_\omega \lambda(\omega) X(\omega) > 0.$

2. For any predictable ϕ

$$\sum_\omega \lambda(\omega) \tilde{G}_N(\phi)(\omega) = 0.$$

From Property 1. we deduce that $\lambda(\omega) > 0$ for all $\omega \in \Omega$, so that the probability $\mathbf{P}^*$ defined by

$$\mathbf{P}^*(\{\omega\}) = \frac{\lambda(\omega)}{\sum_{\omega' \in \Omega} \lambda(\omega')}$$

is equivalent to $\mathbf{P}$.

Moreover, if we denote by $\mathbf{E}^*$ the expectation under measure $\mathbf{P}^*$, Property 2. means that, for any predictable process (ϕ_n) in $\mathbb{R}^d$,

$$\mathbf{E}^* \left(\sum_{j=1}^N \phi_j \Delta \tilde{S}_j \right) = 0.$$

It follows that for all $i \in \{1, \ldots, d\}$ and any predictable sequence (ϕ_n^i) in $\mathbb{R}$, we have

$$\mathbf{E}^* \left(\sum_{j=1}^{N} \phi_j^i \Delta \tilde{S}_j^i \right) = 0.$$

Therefore, according to Proposition 1.2.4, we conclude that the discounted prices $(\tilde{S}_n^1), \ldots, (\tilde{S}_n^d)$ are $\mathbf{P}^*$ martingales. $\qquad\square$

1.3 Complete markets and option pricing

1.3.1 Complete markets

We shall define a *European option** of maturity N by giving its payoff $h \geq 0$, $\mathcal{F}_N$-measurable. For instance, a *call* on the underlying S^1 with strike price K will be defined by setting: $h = \left(S_N^1 - K \right)_+$. A *put* on the same underlying asset with the same strike price K will be defined by $h = \left(K - S_N^1 \right)_+$. In those two examples, which are actually the two most important in practice, h is a function of S_N only. There are some options dependent on the whole path of the underlying asset, i.e. h is a function of $S_0, S_1, \ldots, S_N$. That is the case of the so-called *Asian options* where the strike price is equal to the average of the stock prices observed during a certain period of time before maturity.

Definition 1.3.1 *The contingent claim defined by h is attainable if there exists an admissible strategy worth h at time N.*

Remark 1.3.2 In a viable financial market, we just need to find a *self-financing* strategy worth h at maturity to say that h is attainable. Indeed, if ϕ is a self-financing strategy and if $\mathbf{P}^*$ is a probability measure equivalent to P under which discounted prices are martingales, then $\left(\tilde{V}_n(\phi) \right)$ is also a $\mathbf{P}^*$-martingale, being a martingale transform. Hence, for $n \in \{0, \ldots, N\}$ $\tilde{V}_n(\phi) = \mathbf{E}^* \left(\tilde{V}_N(\phi) | \mathcal{F}_n \right)$. Clearly, if $\tilde{V}_N(\phi) \geq 0$ (in particular if $V_N(\phi) = h$), the strategy ϕ is admissible.

Definition 1.3.3 *The market is* complete *if every contingent claim is attainable.*

To assume that a financial market is complete is a rather restrictive assumption that does not have such a clear economic justification as the no-arbitrage assumption. The interest of complete markets is that it allows us to derive a simple theory of contingent claim pricing and hedging. The Cox-Ross-Rubinstein model, that we shall study in the next section, is a very simple example of complete market modelling. The following theorem gives a precise characterisation of complete, viable financial markets.

---

* Or more generally a contingent claim.

Theorem 1.3.4 *A viable market is complete if and only if there exists a unique probability measure* $\mathbf{P}^*$ *equivalent to* $\mathbf{P}$ *under which discounted prices are martingales.*

The probability $\mathbf{P}^*$ will appear to be the *computing tool* whereby we can derive closed-form pricing formulae and hedging strategies.

Proof. (a) Let us assume that the market is viable and complete. Then, any non-negative, $\mathcal{F}_N$-measurable random variable h can be written as $h = V_N(\phi)$, where ϕ is an admissible strategy that replicates the contingent claim h. Since ϕ is self-financing, we know that

$$\frac{h}{S_N^0} = \tilde{V}_N(\phi) = V_0(\phi) + \sum_{j=1}^{N} \phi_j . \Delta \tilde{S}_j.$$

Thus, if $\mathbf{P}_1$ and $\mathbf{P}_2$ are two probability measures under which discounted prices are martingales, $\left(\tilde{V}_n(\phi) \right)_{0 \le n \le N}$ is a martingale under both $\mathbf{P}_1$ and $\mathbf{P}_2$. It follows that, for $i = 1$ or $i = 2$

$$\mathbf{E}_i \left(\tilde{V}_N(\phi) \right) = \mathbf{E}_i (V_0(\phi)) = V_0(\phi),$$

the last equality coming from the fact that $\mathcal{F}_0 = \{ \emptyset, \Omega \}$. Therefore

$$\mathbf{E}_1 \left(\frac{h}{S_N^0} \right) = \mathbf{E}_2 \left(\frac{h}{S_N^0} \right)$$

and, since h is arbitrary, $\mathbf{P}_1 = \mathbf{P}_2$ on the whole σ-algebra $\mathcal{F}_N$ assumed to be equal to $\mathcal{F}$.

(b) Let us assume that the market is viable and incomplete. Then, there exists a random variable $h \ge 0$ which is not attainable. We call $\tilde{\mathcal{V}}$ the set of random variables of the form

$$U_0 + \sum_{n=1}^{N} \phi_n . \Delta \tilde{S}_n, \tag{1.1}$$

where U_0 is $\mathcal{F}_0$-measurable and $\left((\phi_n^1, \ldots, \phi_n^d) \right)_{0 \le n \le N}$ is an $\mathbb{R}^d$-valued predictable process.

It follows from Proposition 1.1.3 and Remark 1.3.2 that the variable h/S_N^0 does not belong to $\tilde{\mathcal{V}}$. Hence, $\tilde{\mathcal{V}}$ is a strict subset of the set of all random variables on $(\Omega, \mathcal{F})$. Therefore, if $\mathbf{P}^*$ is a probability equivalent to $\mathbf{P}$ under which discounted prices are martingales and if we define the following scalar product on the set of random variables $(X, Y) \mapsto \mathbf{E}^* (XY)$, we notice that there exists a non-zero random variable X orthogonal to $\tilde{\mathcal{V}}$. We also write

$$\mathbf{P}^{**} (\{\omega\}) = \left(1 + \frac{X(\omega)}{2\|X\|_\infty} \right) \mathbf{P}^* (\{\omega\})$$

with $\|X\|_\infty = \sup_{\omega \in \Omega} |X(\omega)|$. Because $\mathbf{E}^*(X) = 0$, that defines a new probability measure equivalent to $\mathbf{P}$ and different from $\mathbf{P}^*$. Moreover

$$\mathbf{E}^{**}\left(\sum_{n=1}^N \phi_n . \Delta \tilde{S}_n\right) = 0$$

for any predictable process $\left((\phi_n^1, \ldots, \phi_n^d)\right)_{0 \le n \le N}$. It follows from Proposition 1.2.4 that $(\tilde{S}_n)_{0 \le n \le N}$ is a $\mathbf{P}^{**}$-martingale.   □

1.3.2 Pricing and hedging contingent claims in complete markets

The market is assumed to be viable and complete and we denote by $\mathbf{P}^*$ the unique probability measure under which the discounted prices of financial assets are martingales. Let h be an $\mathcal{F}_N$-measurable, non-negative random variable and ϕ be an admissible strategy replicating the contingent claim hence defined, i.e.

$$V_N(\phi) = h.$$

The sequence $\left(\tilde{V}_n\right)_{0 \le n \le N}$ is a $\mathbf{P}^*$-martingale, and consequently

$$V_0(\phi) = \mathbf{E}^*\left(\tilde{V}_N(\phi)\right),$$

that is $V_0(\phi) = \mathbf{E}^*\left(h/S_N^0\right)$ and more generally

$$V_n(\phi) = S_n^0 \mathbf{E}^*\left(\frac{h}{S_N^0} | \mathcal{F}_n\right), \quad n = 0, 1, \ldots, N.$$

At any time, the value of an admissible strategy replicating h is completely determined by h. It seems quite natural to call $V_n(\phi)$ the price of the option: that is the wealth needed at time n to replicate h at time N by following the strategy ϕ. If, at time 0, an investor sells the option for

$$\mathbf{E}^*\left(\frac{h}{S_N^0}\right),$$

he can follow a replicating strategy ϕ in order to generate an amount h at time N. In other words, the investor is *perfectly hedged*.

Remark 1.3.5 It is important to notice that the computation of the option price only requires the knowledge of $\mathbf{P}^*$ and not $\mathbf{P}$. We could have just considered a *measurable* space $(\Omega, \mathcal{F})$ equipped with the filtration $(\mathcal{F}_n)$. In other words, we would only define the set of all possible states and the evolution of the information over time. As soon as the probability space and the filtration are specified, we do not need to find the *true* probability of the possible events (say, by statistical means) in order to price the option. The analysis of the Cox-Ross-Rubinstein model will show how we can compute the option price and the hedging strategy in practice.

1.3.3 Introduction to American options

Since an American option can be exercised at any time between 0 and N, we shall define it as a positive sequence (Z_n) adapted to $(\mathcal{F}_n)$, where Z_n is the immediate profit made by exercising the option at time n. In the case of an American option on the stock S^1 with strike price K, $Z_n = \left(S_n^1 - K\right)_+$; in the case of the put, $Z_n = \left(K - S_n^1\right)_+$. In order to define the price of the option associated with $(Z_n)_{0 \leq n \leq N}$, we shall think in terms of a backward induction starting at time N. Indeed, the value of the option at maturity is obviously equal to $U_N = Z_N$. At what price should we sell the option at time $N - 1$? If the holder exercises straight away he will earn Z_{N-1}, or he might exercise at time N in which case the writer must be ready to pay the amount Z_N. Therefore, at time $N - 1$, the writer has to earn the maximum between Z_{N-1} and the amount necessary at time $N - 1$ to generate Z_N at time N. In other words, the writer wants the maximum between Z_{N-1} and the value at time $N - 1$ of an admissible strategy paying off Z_N at time N, i.e. $S_{N-1}^0 \mathbf{E}^* \left(\tilde{Z}_N | \mathcal{F}_{N-1}\right)$, with $\tilde{Z}_N = Z_N / S_N^0$. As we see, it makes sense to price the option at time $N - 1$ as

$$U_{N-1} = \max\left(Z_{N-1}, S_{N-1}^0 \mathbf{E}^* \left(\tilde{Z}_N \middle| \mathcal{F}_{N-1}\right)\right).$$

By induction, we define the American option price for $n = 1, \ldots, N$ by

$$U_{n-1} = \max\left(Z_{n-1}, S_{n-1}^0 \mathbf{E}^* \left(\frac{U_n}{S_n^0} \middle| \mathcal{F}_{n-1}\right)\right).$$

If we assume that the interest rate over one period is constant and equal to r,

$$S_n^0 = (1 + r)^n$$

and

$$U_{n-1} = \max\left(Z_{n-1}, \frac{1}{1+r} \mathbf{E}^* \left(U_n | \mathcal{F}_{n-1}\right)\right),$$

let $\tilde{U}_n = U_n / S_n^0$ be the discounted price of the American option.

Proposition 1.3.6 *The sequence* $\left(\tilde{U}_n\right)_{0 \leq n \leq N}$ *is a* $\mathbf{P}^*$-*supermartingale. It is the smallest* $\mathbf{P}^*$-*supermartingale that dominates the sequence* $\left(\tilde{Z}_n\right)_{0 \leq n \leq N}$.

We should note that, as opposed to the European case, the discounted price of the American option is generally not a martingale under $\mathbf{P}^*$.

Proof. From the equality

$$\tilde{U}_{n-1} = \max\left(\tilde{Z}_{n-1}, \mathbf{E}^* \left(\tilde{U}_n | \mathcal{F}_{n-1}\right)\right),$$

it follows that $(\tilde{U}_n)_{0 \leq n \leq N}$ is a supermartingale dominating $(\tilde{Z}_n)_{0 \leq n \leq N}$. Let us now consider a supermartingale $(\tilde{T}_n)_{0 \leq n \leq N}$ that dominates $(\tilde{Z}_n)_{0 \leq n \leq N}$. Then $\tilde{T}_N \geq \tilde{U}_N$ and if $\tilde{T}_n \geq \tilde{U}_n$ we have

$$\tilde{T}_{n-1} \geq \mathbf{E}^* \left(\tilde{T}_n | \mathcal{F}_{n-1} \right) \geq \mathbf{E}^* \left(\tilde{U}_n | \mathcal{F}_{n-1} \right)$$

whence

$$\tilde{T}_{n-1} \geq \max \left(\tilde{Z}_{n-1}, \mathbf{E}^* \left(\tilde{U}_n | \mathcal{F}_{n-1} \right) \right) = \tilde{U}_{n-1}.$$

A backward induction proves the assertion that (T_n) dominates $(\tilde{U}_n)$. $\qquad \square$

1.4 Problem: Cox, Ross and Rubinstein model

The Cox-Ross-Rubinstein model is a discrete-time version of the Black-Scholes model. It considers only one risky asset whose price is S_n at time n, $0 \leq n \leq N$, and a riskless asset whose return is r over one period of time. To be consistent with the previous sections, we denote $S_n^0 = (1 + r)^n$.

The risky asset is modelled as follows: between two consecutive periods the relative price change is either a or b, with $-1 < a < b$:

$$S_{n+1} = \begin{cases} S_n(1+a) \\ S_n(1+b). \end{cases}$$

The initial stock price S_0 is given. The set of possible states is then $\Omega = \{1 + a, 1 + b\}^N$. Each N-tuple represents the successive values of the ratio S_{n+1}/S_n, $n = 0, 1, \ldots, N - 1$. We also assume that $\mathcal{F}_0 = \{\emptyset, \Omega\}$ and $\mathcal{F} = \mathcal{P}(\Omega)$. For $n = 1, \ldots, N$, the σ-algebra $\mathcal{F}_n$ is equal to $\sigma(S_1, \ldots, S_n)$ generated by the random variables $S_1, \ldots, S_n$. The assumption that each singleton in Ω has a strictly positive probability implies that $\mathbf{P}$ is defined uniquely up to equivalence. We now introduce the variables $T_n = S_n/S_{n-1}$, for $n = 1, \ldots, N$. If $(x_1, \ldots, x_N)$ is one element of Ω, $\mathbf{P}\{(x_1, \ldots, x_N)\} = \mathbf{P}(T_1 = x_1, \ldots, T_N = x_N)$. As a result, knowing $\mathbf{P}$ is equivalent to knowing the law of the N-tuple $(T_1, T_2, \ldots, T_N)$. We also remark that for $n \geq 1$, $\mathcal{F}_n = \sigma(T_1, \ldots, T_n)$.

1. Show that the discounted price $(\tilde{S}_n)$ is a martingale under $\mathbf{P}$ if and only if $\mathbf{E}(T_{n+1} | \mathcal{F}_n) = 1 + r$, $\forall n \in \{0, 1, \ldots, N - 1\}$.

   The equality $\mathbf{E}(\tilde{S}_{n+1} | \mathcal{F}_n) = \tilde{S}_n$ is equivalent to $\mathbf{E}(\tilde{S}_{n+1}/\tilde{S}_n | \mathcal{F}_n) = 1$, since $\tilde{S}_n$ is $\mathcal{F}_n$-measurable and this last equality is actually equivalent to $\mathbf{E}(T_{n+1} | \mathcal{F}_n) = 1 + r$.

2. Deduce that r must belong to $]a, b[$ for the market to be arbitrage-free.

   If the market is viable, there exists a probability $\mathbf{P}^*$ equivalent to $\mathbf{P}$, under which $(\tilde{S}_n)$ is a martingale. Thus, according to Question 1.

   $$\mathbf{E}^*(T_{n+1} | \mathcal{F}_n) = 1 + r$$

   and therefore $\mathbf{E}^*(T_{n+1}) = 1 + r$. Since T_{n+1} is either equal to $1 + a$ or $1 + b$ with non-zero probability, we necessarily have $(1 + r) \in]1 + a, 1 + b[$.

3. Give examples of arbitrage strategies if the no-arbitrage condition derived in Question (2.) is not satisfied.

   Assume for instance that $r \leq a$. By borrowing an amount S_0 at time 0, we can purchase one share of the risky asset. At time N, we pay the loan back and sell the risky asset. We realised a profit equal to $S_N - S_0(1+r)^N$ which is always positive, since $S_N \geq S_0(1+a)^N$. Moreover, it is strictly positive with non-zero probability. There is arbitrage opportunity. If $r \geq b$ we can make a riskless profit by short-selling the risky asset.

4. From now on, we assume that $r \in]a, b[$ and we write $p = (b-r)/(b-a)$. Show that $(\tilde{S}_n)$ is a **P**-martingale if and only if the random variables $T_1, T_2, \ldots, T_N$ are independent, identically distributed (IID) *and* their distribution is given by: $\mathbf{P}(T_1 = 1+a) = p = 1 - \mathbf{P}(T_1 = 1+b)$. Conclude that the market is arbitrage-free and complete.

   If T_i are independent and satisfy $\mathbf{P}(T_i = 1+a) = p = 1 - \mathbf{P}(T_i = 1+b)$, we have
   $$\mathbf{E}(T_{n+1}|\mathcal{F}_n) = \mathbf{E}(T_{n+1}) = p(1+a) + (1-p)(1+b) = 1+r$$
   and thus, $(\tilde{S}_n)$ is a **P**-martingale, according to Question 1.

   Conversely, if for $n = 0, 1, \ldots, N-1$, $\mathbf{E}(T_{n+1}|\mathcal{F}_n) = 1+r$, we can write
   $$(1+a)\mathbf{E}\left(\mathbf{1}_{\{T_{n+1}=1+a\}}|\mathcal{F}_n\right) + (1+b)\mathbf{E}\left(\mathbf{1}_{\{T_{n+1}=1+b\}}|\mathcal{F}_n\right) = 1+r.$$
   Then, the following equality
   $$\mathbf{E}\left(\mathbf{1}_{\{T_{n+1}=1+a\}}|\mathcal{F}_n\right) + \mathbf{E}\left(\mathbf{1}_{\{T_{n+1}=1+b\}}|\mathcal{F}_n\right) = 1,$$
   implies that $\mathbf{E}\left(\mathbf{1}_{\{T_{n+1}=1+a\}}|\mathcal{F}_n\right) = p$ and $\mathbf{E}\left(\mathbf{1}_{\{T_{n+1}=1+b\}}|\mathcal{F}_n\right) = 1-p$. By induction, we prove that for any $x_i \in \{1+a, 1+b\}$,
   $$\mathbf{P}\left(T_1 = x_1, \ldots, T_n = x_n\right) = \prod_{i=1}^{n} p_i$$
   where $p_i = p$ if $x_i = 1+a$ and $p_i = 1-p$ if $x_i = 1+b$. That shows that the variables T_i are IID under measure **P** and that $\mathbf{P}(T_i = 1+a) = p$.

   We have shown that the very fact that $(\tilde{S}_n)$ is a **P**-martingale uniquely determines the distribution of the N-tuple $(T_1, T_2, \ldots, T_N)$ under **P**, hence the measure **P** itself. Therefore, the market is arbitrage-free and complete.

5. We denote by C_n (resp. P_n) the value at time n, of a European call (resp. put) on a share of stock, with strike price K and maturity N.

   (a) Derive the put/call parity equation
   $$C_n - P_n = S_n - K(1+r)^{-(N-n)},$$
   knowing the put/call prices in their conditional expectation form.

   If we denote $\mathbf{E}^*$ the expectation with respect to the probability measure $\mathbf{P}^*$ under which $(\tilde{S}_n)$ is a martingale, we have
   $$\begin{aligned} C_n - P_n &= (1+r)^{-(N-n)}\mathbf{E}^* \left((S_N - K)_+ - (K - S_N)_+|\mathcal{F}_n\right) \\ &= (1+r)^{-(N-n)}\mathbf{E}^* \left(S_N - K|\mathcal{F}_n\right) \\ &= S_n - K(1+r)^{-(N-n)}, \end{aligned}$$

the last equality comes from the fact that $(\tilde{S}_n)$ is a $\mathbf{P}^*$-martingale.

(b)   Show that we can write $C_n = c(n, S_n)$ where c is a function of K, a, b, r and p.

When we write $S_N = S_n \prod_{i=n+1}^{N} T_i$, we get

$$C_n = (1+r)^{-(N-n)} \mathbf{E}^* \left(\left(S_n \prod_{i=n+1}^{N} T_i - K \right)_+ \Bigg| \mathcal{F}_n \right).$$

Since under the probability $\mathbf{P}^*$, the random variable $\prod_{i=n+1}^{N} T_i$ is independent of $\mathcal{F}_n$ and since S_n is $\mathcal{F}_n$-measurable, Proposition A.2.5 in the Appendix allows us to write: $C_n = c(n, S_n)$, where c is the function defined by

$$\frac{c(n, x)}{(1+r)^{-(N-n)}}$$

$$= \mathbf{E}^* \left(x \prod_{i=n+1}^{N} T_i - K \right)_+$$

$$= \sum_{j=0}^{N-n} \frac{(N-n)!}{(N-n-j)!j!} p^j (1-p)^{N-n-j} \left(x(1+a)^j (1+b)^{N-n-j} - K \right)_+.$$

6. Show that the replicating strategy of a call is characterised by a quantity $H_n = \Delta(n, S_{n-1})$ at time n, where Δ will be expressed in terms of function c.

We denote H_n^0 the number of riskless assets in the replicating portfolio. We have

$$H_n^0(1+r)^n + H_n S_n = c(n, S_n).$$

Since H_n^0 and H_n are $\mathcal{F}_{n-1}$-measurable, they are functions of $S_1, \dots, S_{n-1}$ only and, since S_n is equal to $S_{n-1}(1+a)$ or $S_{n-1}(1+b)$, the previous equality implies

$$H_n^0(1+r)^n + H_n S_{n-1}(1+a) = c(n, S_{n-1}(1+a))$$

and

$$H_n^0(1+r)^n + H_n S_{n-1}(1+b) = c(n, S_{n-1}(1+b)).$$

Subtracting one from the other, it turns out that

$$\Delta(n, x) = \frac{c(n, x(1+b)) - c(n, x(1+a))}{x(b-a)}.$$

7. We can now use the model to price a call or a put with maturity T on a single stock. In order to do that, we study the asymptotic case when N converges to infinity, and $r = RT/N$, $\log((1+a)/(1+r)) = -\sigma/\sqrt{N}$ and $\log((1+b)/(1+r)) = \sigma/\sqrt{N}$. The real number R is interpreted as the *instantaneous rate* at all times between 0 and T, because $e^{RT} = \lim_{N\to\infty}(1+r)^N$. σ^2 can be seen as the limit variance, under measure $\mathbf{P}^*$, of the variable $\log(S_N)$, when N converges to infinity.

(a) Let $(Y_N)_{N\geq 1}$ be a sequence of random variables equal to

$$Y_N = X_1^N + X_2^N + \cdots + X_N^N$$

where, for each N, the random variables X_i^N are IID, belong to

$$\{-\sigma/\sqrt{N}, \sigma/\sqrt{N}\},$$

and their mean is equal to μ_N, with $\lim_{N\to\infty}(N\mu_N) = \mu$. Show that the sequence (Y_N) converges in law towards a Gaussian variable with mean μ and variance σ^2.

We just need to study the convergence of the characteristic function ϕ_{Y_N} of Y_N. We obtain

$$
\begin{aligned}
\phi_{Y_N}(u) = \mathbf{E}\left(\exp(iuY_N)\right) &= \prod_{j=1}^{N} \mathbf{E}\left(\exp\left(iuX_j^N\right)\right) \\
&= \left(\mathbf{E}\left(\exp\left(iuX_1^N\right)\right)\right)^N \\
&= \left(1 + iu\mu_N - \sigma^2 u^2/2N + o(1/N)\right)^N.
\end{aligned}
$$

Hence, $\lim_{N\to\infty} \phi_{Y_N}(u) = \exp\left(iu\mu - \sigma^2 u^2/2\right)$, which proves the convergence in law.

(b) Give explicitly the asymptotic prices of the put and the call at time 0.

For a certain N, the put price at time 0 is given by

$$
\begin{aligned}
P_0^{(N)} &= (1 + RT/N)^{-N} \mathbf{E}^*\left(K - S_0 \prod_{n=1}^{N} T_n\right)_+ \\
&= \mathbf{E}^*\left((1 + RT/N)^{-N} K - S_0 \exp(Y_N)\right)_+
\end{aligned}
$$

with $Y_N = \sum_{n=1}^{N} \log(T_n/(1 + r))$. According to the assumptions, the variables $X_j^N = \log(T_j/(1 + r))$ are valued in $\{-\sigma/\sqrt{N}, \sigma/\sqrt{N}\}$, and are IID under probability $\mathbf{P}^*$. Moreover

$$\mathbf{E}^*(X_j^N) = (1 - 2p)\frac{\sigma}{\sqrt{N}} = \frac{2 - e^{\sigma/\sqrt{N}} - e^{-\sigma/\sqrt{N}}}{e^{\sigma/\sqrt{N}} - e^{-\sigma/\sqrt{N}}} \frac{\sigma}{\sqrt{N}}.$$

Therefore, the sequence (Y_N) satisfies the conditions of Question 7.(a), with $\mu = -\sigma^2/2$. If we write $\psi(y) = (Ke^{-RT} - S_0 e^y)_+$, we are able to write

$$
\begin{aligned}
|P_0^{(N)} &- \mathbf{E}^*\left(\psi(Y_N)\right)| \\
&= \left|\mathbf{E}^*\left(\left((1 + RT/N)^{-N} K - S_0 \exp(Y_N)\right)_+ \right.\right. \\
&\quad \left.\left. - \left(Ke^{-RT} - S_0 \exp(Y_N)\right)_+\right)\right| \\
&\leq K\left|(1 + RT/N)^{-N} - e^{-RT}\right|.
\end{aligned}
$$

Since ψ is a bounded [†], continuous function and because the sequence (Y_N) converges in law, we conclude that

$$\lim_{N\to\infty} P_0^{(N)} = \lim_{N\to\infty} \mathbf{E}^*\left(\psi(Y_N)\right)$$

[†] It is precisely to be able to work with a bounded function that we studied the put first.

$$= \frac{1}{\sqrt{2\pi}} \int_{-\infty}^{+\infty} (Ke^{-RT} - S_0 e^{-\sigma^2/2 + \sigma y})_+ e^{-y^2/2} dy.$$

The integral can be expressed easily in terms of the cumulative normal distribution F, so that

$$\lim_{N \to \infty} P_0^{(N)} = Ke^{-RT} F(-d_2) - S_0 F(-d_1),$$

where $d_1 = (\log(x/K) + RT + \sigma^2/2)/\sigma$, $d_2 = d_1 - \sigma$ and

$$F(d) = \frac{1}{\sqrt{2\pi}} \int_{-\infty}^{d} e^{-x^2/2} dx.$$

The price of the call follows easily from put/call parity $\lim_{N \to \infty} C_0^{(N)} = S_0 F(d_1) - Ke^{-RT} F(d_2)$.

Remark 1.4.1 We note that the only non-directly observable parameter is σ. Its interpretation as a variance suggests that it should be estimated by statistical methods. However, we shall tackle this question in Chapter 4.

Notes: We have assumed throughout this chapter that the risky assets were not offering any dividend. Actually, Huang and Litzenberger (1988) apply the same ideas to answer the same questions when the stock is carrying dividends. The theorem of characterisation of complete markets can also be proved with infinite probability spaces (cf. Dalang, Morton and Willinger (1990) and Morton (1989)). In continuous time, the problem is much more tricky (cf. Harrison and Kreps (1979), Stricker (1990) and Delbaen and Schachermayer (1994)). The theory of complete markets in continuous-time was developed by Harrison and Pliska (1981, 1983). An elementary presentation of the Cox-Ross-Rubinstein model is given in the book by J.C. Cox and M. Rubinstein (1985).

2

Optimal stopping problem and American options

The purpose of this chapter is to address the pricing and hedging of American options and to establish the link between these questions and the optimal stopping problem. To do so, we will need to define the notion of optimal stopping time, which will enable us to model exercise strategies for American options. We will also define the Snell envelope, which is the fundamental concept used to solve the optimal stopping problem. The application of these concepts to American options will be described in Section 2.5.

2.1 Stopping time

The buyer of an American option can exercise its right at any time until maturity. The decision to exercise or not at time n will be made according to the information available at time n. In a discrete-time model built on a finite filtered space $(\Omega, \mathcal{F}, (\mathcal{F}_n)_{0 \leq n \leq N}, \mathbf{P})$, the exercise date is described by a random variable called stopping time.

Definition 2.1.1 *A random variable ν taking values in $\{0, 1, 2, \ldots, N\}$ is a stopping time if, for any $n \in \{0, 1, \cdots, N\}$,*

$$\{\nu = n\} \in \mathcal{F}_n.$$

Remark 2.1.2 As in the previous chapter, we assume that $\mathcal{F} = \mathcal{P}(\Omega)$ and $\mathbf{P}(\{\omega\}) > 0$, $\forall \omega \in \Omega$. This hypothesis is nonetheless not essential: if it does not hold, the results presented in this chapter remain true almost surely. However, we will not assume $\mathcal{F}_0 = \{\emptyset, \Omega\}$ and $\mathcal{F}_N = \mathcal{F}$, except in Section 2.5, dedicated to finance.

Remark 2.1.3 The reader can verify, as an exercise, that ν is a stopping time if and only if, for any $n \in \{0, 1, \ldots, N\}$,

$$\{\nu \leq n\} \in \mathcal{F}_n.$$

We will use this equivalent definition to generalise the concept of stopping time to the continuous-time setting.

Let us introduce now the concept of a 'sequence stopped at a stopping time'. Let $(X_n)_{0 \leq n \leq N}$ be a sequence adapted to the filtration $(\mathcal{F}_n)_{0 \leq n \leq N}$ and let ν be a stopping time. The sequence stopped at time ν is defined as

$$X_n^\nu(\omega) = X_{\nu(\omega) \wedge n}(\omega),$$

i.e., on the set $\{\nu = j\}$ we have

$$X_n^\nu = \begin{cases} X_j & \text{if } j \leq n \\ X_n & \text{if } j > n. \end{cases}$$

Note that $X_N^\nu(\omega) = X_{\nu(\omega)}(\omega) \ (= X_j \text{ on } \{\nu = j\})$.

Proposition 2.1.4 *Let (X_n) be an adapted sequence and ν be a stopping time. The stopped sequence $(X_n^\nu)_{0 \leq n \leq N}$ is adapted. Moreover, if (X_n) is a martingale (resp. a supermartingale), then (X_n^ν) is a martingale (resp. a supermartingale).*

Proof. We see that, for $n \geq 1$, we have

$$X_{\nu \wedge n} = X_0 + \sum_{j=1}^{n} \phi_j (X_j - X_{j-1}),$$

where $\phi_j = 1_{\{j \leq \nu\}}$. Since $\{j \leq \nu\}$ is the complement of the set $\{\nu < j\} = \{\nu \leq j - 1\}$, the process $(\phi_n)_{0 \leq n \leq N}$ is predictable.

It is clear then that $(X_{\nu \wedge n})_{0 \leq n \leq N}$ is adapted to the filtration $(\mathcal{F}_n)_{0 \leq n \leq N}$. Furthermore, if (X_n) is a martingale, $(X_{\nu \wedge n})$ is also a martingale with respect to $(\mathcal{F}_n)$, since it is the martingale transform of (X_n). Similarly, we can show that if the sequence (X_n) is a supermartingale (resp. a submartingale), the stopped sequence is still a supermartingale (resp. a submartingale) using the predictability and the non-negativity of $(\phi_j)_{0 \leq j \leq N}$. $\square$

2.2 The Snell envelope

In this section, we consider an adapted sequence $(Z_n)_{0 \leq n \leq N}$, and define the sequence $(U_n)_{0 \leq n \leq N}$ as follows:

$$\begin{cases} U_N &= Z_N \\ U_n &= \max\left(Z_n, \mathbf{E}\left(U_{n+1} | \mathcal{F}_n\right)\right) \quad \forall n \leq N - 1. \end{cases}$$

The study of this sequence is motivated by our first approach of American options (Section 1.3.3 of Chapter 1). We already know, by Proposition 1.3.6 of Chapter 1, that $(U_n)_{0 \leq n \leq N}$ is the smallest supermartingale that dominates the sequence $(Z_n)_{0 \leq n \leq N}$. We call it the Snell envelope of the sequence $(Z_n)_{0 \leq n \leq N}$.

By definition, U_n is greater than Z_n (with equality for $n = N$) and in the case of a strict inequality, $U_n = \mathbf{E}(U_{n+1} | \mathcal{F}_n)$. It suggests that, by stopping adequately the sequence (U_n), it is possible to obtain a martingale, as the following proposition shows.

Proposition 2.2.1 *The random variable defined by*

$$\nu_0 = \inf \{n \geq 0 | U_n = Z_n\} \tag{2.1}$$

is a stopping time and the stopped sequence $(U_{n \wedge \nu_0})_{0 \leq n \leq N}$ is a martingale.

Proof. Since $U_N = Z_N$, ν_0 is a well-defined element of $\{0, 1, \ldots, N\}$ and we have

$$\{\nu_0 = 0\} = \{U_0 = Z_0\} \in \mathcal{F}_0,$$

and for $k \geq 1$

$$\{\nu_0 = k\} = \{U_0 > Z_0\} \cap \cdots \cap \{U_{k-1} > Z_{k-1}\} \cap \{U_k = Z_k\} \in \mathcal{F}_k.$$

To demonstrate that $(U_n^{\nu_0})$ is a martingale, we write as in the proof of Proposition 2.1.4:

$$U_n^{\nu_0} = U_{n \wedge \nu_0} = U_0 + \sum_{j=1}^{n} \phi_j \Delta U_j,$$

where $\phi_j = 1_{\{\nu_0 \geq j\}}$. So that, for $n \in \{0, 1, \ldots, N-1\}$,

$$
\begin{aligned}
U_{n+1}^{\nu_0} - U_n^{\nu_0} &= \phi_{n+1} (U_{n+1} - U_n) \\
&= 1_{\{n+1 \leq \nu_0\}} (U_{n+1} - U_n).
\end{aligned}
$$

By definition, $U_n = \max (Z_n, \mathbf{E}(U_{n+1} | \mathcal{F}_n))$ and on the set $\{n+1 \leq \nu_0\}$, $U_n > Z_n$. Consequently $U_n = \mathbf{E}(U_{n+1} | \mathcal{F}_n)$ and we deduce

$$U_{n+1}^{\nu_0} - U_n^{\nu_0} = 1_{\{n+1 \leq \nu_0\}} (U_{n+1} - \mathbf{E}(U_{n+1} | \mathcal{F}_n))$$

and taking the conditional expectation on both sides of the equality

$$\mathbf{E}\left((U_{n+1}^{\nu_0} - U_n^{\nu_0}) | \mathcal{F}_n \right) = 1_{\{n+1 \leq \nu_0\}} \mathbf{E}\left((U_{n+1} - \mathbf{E}(U_{n+1} | \mathcal{F}_n)) | \mathcal{F}_n \right)$$

because $\{n+1 \leq \nu_0\} \in \mathcal{F}_n$ (since the complement of $\{n+1 \leq \nu_0\}$ is $\{\nu_0 \leq n\}$).

Hence

$$\mathbf{E}\left((U_{n+1}^{\nu_0} - U_n^{\nu_0}) | \mathcal{F}_n \right) = 0,$$

which proves that U^{ν_0} is a martingale. $\qquad\square$

In the remainder, we shall note $\mathcal{T}_{n,N}$ the set of stopping times taking values in $\{n, n+1, \ldots, N\}$. Notice that $\mathcal{T}_{n,N}$ is a finite set since Ω is assumed to be finite. The martingale property of the sequence U^{ν_0} gives the following result which relates the concept of Snell envelope to the optimal stopping problem.

Corollary 2.2.2 *The stopping time ν_0 satisfies*

$$U_0 = \mathbf{E}(Z_{\nu_0} | \mathcal{F}_0) = \sup_{\nu \in \mathcal{T}_{0,N}} \mathbf{E}(Z_\nu | \mathcal{F}_0).$$

If we think of Z_n as the total winnings of a gambler after n games, we see that stopping at time ν_0 maximises the expected gain given $\mathcal{F}_0$.

Proof. Since U^{ν_0} is a martingale, we have

$$U_0 = U_0^{\nu_0} = \mathbf{E}(U_N^{\nu_0} | \mathcal{F}_0) = \mathbf{E}(U_{\nu_0} | \mathcal{F}_0) = \mathbf{E}(Z_{\nu_0} | \mathcal{F}_0).$$

On the other hand, if $\nu \in \mathcal{T}_{0,N}$, the stopped sequence U^ν is a supermartingale. So that

$$
\begin{aligned}
U_0 &\geq \mathbf{E}\,(U_N^\nu|\mathcal{F}_0) = \mathbf{E}\,(U_\nu|\mathcal{F}_0) \\
&\geq \mathbf{E}\,(Z_\nu|\mathcal{F}_0)\,,
\end{aligned}
$$

which yields the result. $\square$

Remark 2.2.3 An immediate generalisation of Corollary 2.2.2 gives

$$
\begin{aligned}
U_n &= \sup_{\nu \in \mathcal{T}_{n,N}} \mathbf{E}\,(Z_\nu|\mathcal{F}_n) \\
&= \mathbf{E}\,(Z_{\nu_n}|\mathcal{F}_n)\,,
\end{aligned}
$$

where $\nu_n = \inf\{j \geq n | U_j = Z_j\}$.

Definition 2.2.4 *A stopping time ν is called optimal for the sequence $(Z_n)_{0 \leq n \leq N}$ if*

$$
\mathbf{E}\,(Z_\nu|\mathcal{F}_0) = \sup_{\mathcal{T}_{0,N}} \mathbf{E}\,(Z_\nu|\mathcal{F}_0)\,.
$$

We can see that ν_0 is optimal. The following result gives a characterisation of optimal stopping times that shows that ν_0 is the smallest optimal stopping time.

Theorem 2.2.5 *A stopping time ν is optimal if and only if*

$$
\begin{cases}
Z_\nu = U_\nu \\
\text{and } (U_{\nu \wedge n})_{0 \leq n \leq N} \text{ is a martingale.}
\end{cases}
\tag{2.2}
$$

Proof. If the stopped sequence U^ν is a martingale, $U_0 = \mathbf{E}(U_\nu|\mathcal{F}_0)$ and consequently, if (2.2) holds, $U_0 = \mathbf{E}(Z_\nu|\mathcal{F}_0)$. Optimality of ν is then ensured by Corollary 2.2.2.

Conversely, if ν is optimal, we have

$$
U_0 = \mathbf{E}\,(Z_\nu|\mathcal{F}_0) \leq \mathbf{E}\,(U_\nu|\mathcal{F}_0)\,.
$$

But, since U^ν is a supermartingale,

$$
\mathbf{E}\,(U_\nu|\mathcal{F}_0) \leq U_0.
$$

Therefore

$$
\mathbf{E}\,(U_\nu|\mathcal{F}_0) = \mathbf{E}\,(Z_\nu|\mathcal{F}_0)
$$

and since $U_\nu \geq Z_\nu$, $U_\nu = Z_\nu$.

Since $\mathbf{E}\,(U_\nu|\mathcal{F}_0) = U_0$ and from the following inequalities

$$
U_0 \geq \mathbf{E}\,(U_{\nu \wedge n}|\mathcal{F}_0) \geq \mathbf{E}\,(U_\nu|\mathcal{F}_0)
$$

(based on the supermartingale property of (U_n^ν)) we get

$$
\mathbf{E}\,(U_{\nu \wedge n}|\mathcal{F}_0) = \mathbf{E}\,(U_\nu|\mathcal{F}_0) = \mathbf{E}\,(\mathbf{E}\,(U_\nu|\mathcal{F}_n)|\,\mathcal{F}_0)\,.
$$

But we have $U_{\nu \wedge n} \geq \mathbf{E}\,(U_\nu|\mathcal{F}_n)$, therefore $U_{\nu \wedge n} = \mathbf{E}\,(U_\nu|\mathcal{F}_n)$, which proves that (U_n^ν) is a martingale. $\square$

2.3 Decomposition of supermartingales

The following decomposition (commonly called 'Doob decomposition') is used in viable complete market models to associate any supermartingale with a trading strategy for which consumption is allowed (see Exercise 5 for that matter).

Proposition 2.3.1 *Every supermartingale* $(U_n)_{0 \leq n \leq N}$ *has the unique following decomposition:*

$$U_n = M_n - A_n,$$

where (M_n) *is a martingale and* (A_n) *is a non-decreasing, predictable process, null at* 0.

Proof. It is clearly seen that the only solution for $n = 0$ is $M_0 = U_0$ and $A_0 = 0$. Then we must have

$$U_{n+1} - U_n = M_{n+1} - M_n - (A_{n+1} - A_n).$$

So that, conditioning both sides with respect to $\mathcal{F}_n$ and using the properties of M and A

$$-(A_{n+1} - A_n) = \mathbf{E}(U_{n+1}|\mathcal{F}_n) - U_n$$

and

$$M_{n+1} - M_n = U_{n+1} - \mathbf{E}(U_{n+1}|\mathcal{F}_n).$$

(M_n) and (A_n) are entirely determined using the previous equations and we see that (M_n) is a martingale and that (A_n) is predictable and non-decreasing (because (U_n) is a supermartingale). $\square$

Suppose then that (U_n) is the Snell envelope of an adapted sequence (Z_n). We can then give a characterisation of the largest optimal stopping time for (Z_n) using the non-decreasing process (A_n) of the Doob decomposition of (U_n):

Proposition 2.3.2 *The largest optimal stopping time for* (Z_n) *is given by*

$$\nu_{\max} = \begin{cases} N & \text{if } A_N = 0 \\ \inf \{n, \ A_{n+1} \neq 0\} & \text{if } A_N \neq 0. \end{cases}$$

Proof. It is straightforward to see that $\nu_{\max}$ is a stopping time using the fact that $(A_n)_{0 \leq n \leq N}$ is predictable. From $U_n = M_n - A_n$ and because $A_j = 0$, for $j \leq \nu_{\max}$, we deduce that $U^{\nu_{\max}} = M^{\nu_{\max}}$ and conclude that $U^{\nu_{\max}}$ is a martingale. To show the optimality of $\nu_{\max}$, it is sufficient to prove

$$U_{\nu_{\max}} = Z_{\nu_{\max}}.$$

We note that

$$U_{\nu_{\max}} = \sum_{j=0}^{N-1} 1_{\{\nu_{\max}=j\}} U_j + 1_{\{\nu_{\max}=N\}} U_N$$

$$= \sum_{j=0}^{N-1} 1_{\{\nu_{\max}=j\}} \max \left(Z_j, \mathbf{E}(U_{j+1}|\mathcal{F}_j) \right) + 1_{\{\nu_{\max}=N\}} Z_N,$$

We have $\mathbf{E}\left(U_{j+1}|\mathcal{F}_j\right) = M_j - A_{j+1}$ and, on the set $\{\nu_{\max} = j\}$, $A_j = 0$ and $A_{j+1} > 0$, so $U_j = M_j$ and $\mathbf{E}\left(U_{j+1}|\mathcal{F}_j\right) = M_j - A_{j+1} < U_j$. It follows that $U_j = \max\left(Z_j, \mathbf{E}\left(U_{j+1}|\mathcal{F}_j\right)\right) = Z_j$. So that finally

$$U_{\nu_{\max}} = Z_{\nu_{\max}}.$$

It remains to show that it is the greatest optimal stopping time. If ν is a stopping time such that $\nu \geq \nu_{\max}$ and $\mathbf{P}(\nu > \nu_{\max}) > 0$, then

$$\mathbf{E}(U_\nu) = \mathbf{E}(M_\nu) - \mathbf{E}(A_\nu) = \mathbf{E}(U_0) - \mathbf{E}(A_\nu) < \mathbf{E}(U_0)$$

and U^ν cannot be a martingale, which establishes the claim. $\qquad\square$

2.4 Snell envelope and Markov chains

The aim of this section is to compute Snell envelopes in a Markovian setting. A sequence $(X_n)_{n\geq 0}$ of random variables taking their values in a finite set E is called a Markov chain if, for any integer $n \geq 1$ and any elements $x_0, x_1, \ldots, x_{n-1}, x, y$ of E, we have

$$\mathbf{P}(X_{n+1} = y|X_0 = x_0, \ldots, X_{n-1} = x_{n-1}, X_n = x) = \mathbf{P}(X_{n+1} = y|X_n = x).$$

The chain is said to be homogeneous if the value $P(x,y) = \mathbf{P}\left(X_{n+1} = y|X_n = x\right)$ does not depend on n. The matrix $P = (P(x,y))_{(x,y)\in E\times E}$, indexed by $E \times E$, is then called the transition matrix of the chain. The matrix P has non-negative entries and satisfies: $\sum_{y\in E} P(x,y) = 1$ for all $x \in E$; it is said to be a stochastic matrix. On a filtered probability space $\left(\Omega, \mathcal{F}, (\mathcal{F}_n)_{0\leq n\leq N}, \mathbf{P}\right)$, we can define the notion of a Markov chain with respect to the filtration:

Definition 2.4.1 *A sequence $(X_n)_{0\leq n\leq N}$ of random variables taking values in E is a homogeneous Markov chain with respect to the filtration $(\mathcal{F}_n)_{0\leq n\leq N}$, with transition matrix P, if (X_n) is adapted and if for any real-valued function f on E, we have*

$$\mathbf{E}\left(f\left(X_{n+1}\right)|\mathcal{F}_n\right) = Pf\left(X_n\right),$$

where Pf represents the function which maps $x \in E$ to $Pf(x) = \sum_{y\in E} P(x,y)f(y)$.

Note that, if one interprets real-valued functions on E as matrices with a single column indexed by E, then Pf is indeed the product of the two matrices P and f. It can also be easily seen that a Markov chain, as defined at the beginning of the section, is a Markov chain with respect to its natural filtration, defined by $\mathcal{F}_n = \sigma(X_0, \ldots, X_n)$.

The following proposition is an immediate consequence of the latter definition and the definition of a Snell envelope.

Proposition 2.4.2 *Let (Z_n) be an adapted sequence defined by $Z_n = \psi(n, X_n)$, where (X_n) is a homogeneous Markov chain with transition matrix P, taking values in E, and ψ is a function from $\mathbf{N} \times E$ to $\mathbb{R}$. Then, the Snell envelope (U_n)*

of the sequence (Z_n) is given by $U_n = u(n, X_n)$, where the function u is defined by

$$u(N, x) = \psi(N, x) \quad \forall x \in E$$

and, for $n \leq N - 1$,

$$u(n, \cdot) = \max\left(\psi(n, \cdot), Pu(n + 1, \cdot)\right).$$

2.5 Application to American options

From now on, we will work in a viable complete market. The modelling will be based on the filtered space $\left(\Omega, \mathcal{F}, (\mathcal{F}_n)_{0 \leq n \leq N}, \mathbf{P}\right)$ and, as in Sections 1.3.1 and 1.3.3 of Chapter 1, we will denote by $\mathbf{P}^*$ the unique probability under which the discounted asset prices are martingales.

2.5.1 Hedging American options

In Section 1.3.3 of Chapter 1, we defined the value process (U_n) of an American option described by the sequence (Z_n), by the system

$$\begin{cases} U_N &= Z_N \\ U_n &= \max\left(Z_n, S_n^0 \mathbf{E}^* \left(U_{n+1}/S_{n+1}^0 | \mathcal{F}_n\right)\right) \quad \forall n \leq N - 1. \end{cases}$$

Thus, the sequence $(\tilde{U}_n)$ defined by $\tilde{U}_n = U_n/S_n^0$ (discounted price of the option) is the Snell envelope, under P^*, of the sequence $(\tilde{Z}_n)$. We deduce from the above Section 2.2 that

$$\tilde{U}_n = \sup_{\nu \in \mathcal{T}_{n,N}} \mathbf{E}^* \left(\tilde{Z}_\nu | \mathcal{F}_n\right)$$

and consequently

$$U_n = S_n^0 \sup_{\nu \in \mathcal{T}_{n,N}} \mathbf{E}^* \left(\frac{Z_\nu}{S_\nu^0} | \mathcal{F}_n\right).$$

From Section 2.3, we can write

$$\tilde{U}_n = \tilde{M}_n - \tilde{A}_n,$$

where $(\tilde{M}_n)$ is a $\mathbf{P}^*$-martingale and $(\tilde{A}_n)$ is an increasing predictable process, null at 0. Since the market is complete, there is a self-financing strategy ϕ such that

$$V_N(\phi) = S_N^0 \tilde{M}_N,$$

i.e., $\tilde{V}_N(\phi) = \tilde{M}_N$. For the sequence $\left(\tilde{V}_n(\phi)\right)$ is a $\mathbf{P}^*$-martingale, we have

$$\begin{aligned} \tilde{V}_n(\phi) &= \mathbf{E}^* \left(\tilde{V}_N(\phi) | \mathcal{F}_n\right) \\ &= \mathbf{E}^* \left(\tilde{M}_N | \mathcal{F}_n\right) \\ &= \tilde{M}_n, \end{aligned}$$

and consequently

$$\tilde{U}_n = \tilde{V}_n(\phi) - \tilde{A}_n.$$

Therefore

$$U_n = V_n(\phi) - A_n,$$

where $A_n = S_n^0 \tilde{A}_n$. From the previous equality, it is obvious that the writer of the option can hedge himself perfectly: once he receives the premium $U_0 = V_0(\phi)$, he can generate a wealth equal to $V_n(\phi)$ at time n which is bigger than U_n and *a fortiori* Z_n.

What is the optimal date to exercise the option? The date of exercise is to be chosen among all the stopping times. For the buyer of the option, there is no point in exercising at time n when $U_n > Z_n$, because he would trade an asset worth U_n (the option) for an amount Z_n (by exercising the option). Thus an optimal date τ of exercise is such that $U_\tau = Z_\tau$. On the other hand, there is no point in exercising after the time

$$\nu_{\max} = \inf \{ j, \ A_{j+1} \neq 0 \}$$

(which is equal to $\inf \left\{ j, \ \tilde{A}_{j+1} \neq 0 \right\}$) because, at that time, selling the option provides the holder with a wealth $U_{\nu_{\max}} = V_{\nu_{\max}}(\phi)$ and, following the strategy ϕ from that time, he creates a portfolio whose value is strictly bigger than the option's at times $\nu_{\max} + 1, \nu_{\max} + 2, \ldots, N$. Therefore we set, as a second condition, $\tau \leq \nu_{\max}$, which allows us to say that $\tilde{U}^\tau$ is a martingale. As a result, optimal dates of exercise are optimal stopping times for the sequence $(\tilde{Z}_n)$, under probability $\mathbf{P}^*$. To make this point clear, let us consider the writer's point of view. If he hedges himself using the strategy ϕ as defined above and if the buyer exercises at time τ which is not optimal, then $U_\tau > Z_\tau$ or $A_\tau > 0$. In both cases, the writer makes a profit $V_\tau(\phi) - Z_\tau = U_\tau + A_\tau - Z_\tau$, which is positive.

2.5.2 *American options and European options*

Proposition 2.5.1 *Let C_n be the value at time n of an American option described by an adapted sequence $(Z_n)_{0 \leq n \leq N}$ and let c_n be the value at time n of the European option defined by the $\mathcal{F}_N$-measurable random variable $h = Z_N$. Then, we have $C_n \geq c_n$.*

Moreover, if $c_n \geq Z_n$ for any n, then

$$c_n = C_n \quad \forall n \in \{0, 1, \ldots, N\}.$$

The inequality $C_n \geq c_n$ makes sense since the American option entitles the holder to more rights than its European counterpart.

Proof. For the discounted value $\left(\tilde{C}_n \right)$ is a supermartingale under $\mathbf{P}^*$, we have

$$\tilde{C}_n \geq \mathbf{E}^* \left(\tilde{C}_N | \mathcal{F}_n \right) = \mathbf{E}^* \left(\tilde{c}_N | \mathcal{F}_n \right) = \tilde{c}_n.$$

Hence $C_n \geq c_n$.

If $c_n \geq Z_n$ for any n then the sequence $(\tilde{c}_n)$, which is a martingale under $\mathbf{P}^*$, appears to be a supermartingale (under $\mathbf{P}^*$) and an upper bound of the sequence $(\tilde{Z}_n)$ and consequently

$$\tilde{C}_n \leq \tilde{c}_n \quad \forall n \in \{0, 1, \ldots, N\}.$$

$\square$

Remark 2.5.2 One checks readily that if the relationships of Proposition 2.5.1 did not hold, there would be some arbitrage opportunities by trading the options.

To illustrate the last proposition, let us consider the case of a market with a single risky asset, with price S_n at time n and a constant riskless interest rate, equal to $r \geq 0$ on each period, so that $S_n^0 = (1 + r)^n$. Then, with notations of Proposition 2.5.1, if we take $Z_n = (S_n - K)_+$, c_n is the price at time n of a European call with maturity N and strike price K on one unit of the risky asset and C_n is the price of the corresponding American call. We have

$$\begin{aligned}
\tilde{c}_n &= (1+r)^{-N} \mathbf{E}^* \left((S_N - K)_+ | \mathcal{F}_n \right) \\
&\geq \mathbf{E}^* \left(\tilde{S}_N - K(1+r)^{-N} | \mathcal{F}_n \right) \\
&= \tilde{S}_N - K(1+r)^{-N},
\end{aligned}$$

using the martingale property of $(\tilde{S}_n)$. Hence: $c_n \geq S_n - K(1+r)^{-(N-n)} \geq S_n - K$, for $r \geq 0$. As $c_n \geq 0$, we also have $c_n \geq (S_n - K)_+$ and by Proposition 2.5.1, $C_n = c_n$. There is equality between the price of the European call and the price of the corresponding American call.

This property does not hold for the put, nor in the case of calls on currencies or dividend paying stocks.

Notes: For further discussions on the Snell envelope and optimal stopping, one may consult Neveu (1972), Chapter VI and Dacunha-Castelle and Duflo (1986), Chapter 5, Section 1. For the theory of optimal stopping in the continuous case, see El Karoui (1981) and Shiryayev (1978).

2.6 Exercises

Exercise 1 Let ν be a stopping time with respect to a filtration $(\mathcal{F}_n)_{0 \leq n \leq N}$. We denote by $\mathcal{F}_\nu$ the set of events A such that $A \cap \{\nu = n\} \in \mathcal{F}_n$ for any $n \in \{0, \ldots, N\}$.

1. Show that $\mathcal{F}_\nu$ is a sub-σ-algebra of $\mathcal{F}_N$. $\mathcal{F}_\nu$ is often called 'σ-algebra of events determined prior to the stopping time ν'.

2. Show that the random variable ν is $\mathcal{F}_\nu$-measurable.

3. Let X be a real-valued random variable. Prove the equality

$$\mathbf{E}(X|\mathcal{F}_\nu) = \sum_{j=0}^{N} 1_{\{\nu=j\}} \mathbf{E}(X|\mathcal{F}_j).$$

4. Let τ be a stopping time such that $\tau \geq \nu$. Show that $\mathcal{F}_\nu \subset \mathcal{F}_\tau$.

5. Under the same hypothesis, show that if (M_n) is a martingale, we have

$$M_\nu = \mathbf{E}(M_\tau | \mathcal{F}_\nu).$$

(Hint: first consider the case $\tau = N$.)

Exercise 2 Let (U_n) be the Snell envelope of an adapted sequence (Z_n). Without assuming that $\mathcal{F}_0$ is trivial, show that

$$\mathbf{E}(U_0) = \sup_{\nu \in \mathcal{T}_{0,N}} \mathbf{E}(Z_\nu),$$

and more generally

$$\mathbf{E}(U_n) = \sup_{\nu \in \mathcal{T}_{n,N}} \mathbf{E}(Z_\nu).$$

Exercise 3 Show that ν is optimal according to Definition 2.2.4 if and only if

$$\mathbf{E}(Z_\nu) = \sup_{\tau \in \mathcal{T}_{0,N}} \mathbf{E}(Z_\tau).$$

Exercise 4 The purpose of this exercise is to study the American put in the model of Cox-Ross-Rubinstein. Notations are those of Chapter 1.

1. Show that the price $\mathcal{P}_n$, at time n, of an American put on a share with maturity N and strike price K can be written as

$$\mathcal{P}_n = P_{am}(n, S_n)$$

where $P_{am}(n, x)$ is defined by $P_{am}(N, x) = (K - x)_+$ and, for $n \leq N - 1$,

$$P_{am}(n, x) = \max\left((K - x)_+, \frac{f(n+1, x)}{1+r} \right),$$

with $f(n+1, x) = pP_{am}(n+1, x(1+a)) + (1-p)P_{am}(n+1, x(1+b))$ and $p = (b - r)/(b - a)$.

2. Show that the function $P_{am}(0, .)$ can be expressed as

$$P_{am}(0, x) = \sup_{\nu \in \mathcal{T}_{0,N}} \mathbf{E}^* \left((1+r)^{-\nu}(K - xV_\nu)_+ \right),$$

where the sequence of random variables $(V_n)_{0 \leq n \leq N}$ is defined by: $V_0 = 1$ and, for $n \geq 1$, $V_n = \prod_{i=1}^n U_i$, where the U_i's are some random variables. Give their joint law under $\mathbf{P}^*$.

3. From the last formula, show that the function $x \mapsto P_{am}(0, x)$ is convex and non-increasing.

4. We assume $a < 0$. Show that there is a real number $x^* \in [0, K]$ such that, for $x \leq x^*$, $P_{am}(0, x) = (K - x)_+$ and, for $x \in]x^*, K/(1 + a)^N[$, $P_{am}(0, x) > (K - x)_+$.

5. An agent holds the American put at time 0. For which values of the spot S_0 would he rather exercise his option immediately?

6. Show that the hedging strategy of the American put is determined by a quantity $H_n = \Delta(n, S_{n-1})$ of the risky asset to be held at time n, where Δ can be written as a function of P_{am}.

Exercise 5 Consumption strategies. The self-financing strategies defined in Chapter 1 ruled out any consumption. Consumption strategies can be introduced in the following way: at time n, once the new prices $S_n^0, \ldots, S_n^d$ are quoted, the investor readjusts his positions from ϕ_n to ϕ_{n+1} and selects the wealth γ_{n+1} to be consumed at time $n + 1$. Any endowment being excluded and the new positions being decided given prices at time n, we deduce

$$\phi_{n+1}.S_n = \phi_n.S_n - \gamma_{n+1}. \tag{2.3}$$

So a trading strategy with consumption will be defined as a pair (ϕ, γ), where ϕ is a predictable process taking values in $\mathbb{R}^{d+1}$, representing the numbers of assets held in the portfolio and $\gamma = (\gamma_n)_{1 \leq n \leq N}$ is a predictable process taking values in $\mathbb{R}^+$, representing the wealth consumed at any time. Equation (2.3) gives the relationship between the processes ϕ and γ and replaces the self-financing condition of Chapter 1.

1. Let ϕ be a predictable process taking values in $\mathbb{R}^{d+1}$ and let γ be a predictable process taking values in $\mathbb{R}^+$. We set $V_n(\phi) = \phi_n.S_n$ and $\tilde{V}_n(\phi) = \phi_n.\tilde{S}_n$. Show the equivalence between the following conditions:

   (a) The pair (ϕ, γ) defines a trading strategy with consumption.
   (b) For any $n \in \{1, \ldots, N\}$,

   $$V_n(\phi) = V_0(\phi) + \sum_{j=1}^{n} \phi_j.\Delta S_j - \sum_{j=1}^{n} \gamma_j.$$

   (c) For any $n \in \{1, \ldots, N\}$,

   $$\tilde{V}_n(\phi) = V_0(\phi) + \sum_{j=1}^{n} \phi_j.\Delta \tilde{S}_j - \sum_{j=1}^{n} \gamma_j/S_{j-1}^0.$$

2. In the remainder, we assume that the market is viable and complete and we denote by $\mathbf{P}^*$ the unique probability under which the assets discounted prices are martingales. Show that if the pair (ϕ, γ) defines a trading strategy with consumption, then $(\tilde{V}_n(\phi))$ is a supermartingale under $\mathbf{P}^*$.

3. Let (U_n) be an adapted sequence such that $(\tilde{U}_n)$ is a supermartingale under $\mathbf{P}^*$. Using the Doob decomposition, show that there is a trading strategy with consumption (ϕ, γ) such that $V_n(\phi) = U_n$ for any $n \in \{0, \ldots, N\}$.

4. Let (Z_n) be an adapted sequence. We say that a trading strategy with consumption (ϕ, γ) hedges the American option defined by (Z_n) if $V_n(\phi) \geq Z_n$ for any $n \in \{0, 1, \ldots, N\}$. Show that there is at least one trading strategy with consumption that hedges (Z_n), whose value is precisely the value (U_n) of the American option. Also, prove that any trading strategy with consumption (ϕ, γ) hedging (Z_n) satisfies $V_n(\phi) \geq U_n$, for any $n \in \{0, 1, \ldots, N\}$.

5. Let x be a non-negative number representing the investor's endowment and let $\gamma = (\gamma_n)_{1 \leq n \leq N}$ be a predictable strategy taking values in $\mathbb{R}^+$. The consumption process (γ_n) is said to be budget-feasible from endowment x if there is a predictable process ϕ taking values in $\mathbb{R}^{d+1}$, such that the pair (ϕ, γ) defines a trading strategy with consumption satisfying: $V_0(\phi) = x$ and $V_n(\phi) \geq 0$, for any $n \in \{0, \ldots, N\}$. Show that (γ_n) is budget-feasible from endowment x if and only if $\mathbf{E}^* \left(\sum_{j=1}^{N} \gamma_j / S_{j-1}^0 \right) \leq x$.

3

Brownian motion and stochastic differential equations

The first two chapters of this book were dealing with discrete-time models. We had the opportunity to see the importance of the concepts of martingales, self-financing strategy and Snell envelope. We are going to elaborate on these ideas in a continuous-time framework. In particular, we shall introduce the mathematical tools needed to model financial assets and to price options. In continuous-time, the technical aspects are more advanced and more difficult to handle than in discrete-time, but the main ideas are fundamentally the same.

Why do we consider continuous-time models? The primary motivation comes from the nature of the processes that we want to model. In practice, the price changes in the market are actually so frequent that a discrete-time model can barely follow the moves. On the other hand, continuous-time models lead to more explicit computations, even if numerical methods are sometimes required. Indeed, the most widely used model is the continuous-time Black-Scholes model which leads to an extremely simple formula. As we mentioned in the Introduction, the connections between stochastic processes and finance are not recent. Bachelier (1900), in his dissertation called *Théorie de la spéculation*, is not only among the first to look at the properties of Brownian motion, but he also derived option pricing formulae.

We will be giving a few mathematical definitions in order to understand continuous-time models. In particular, we shall define the Brownian motion since it is the core concept of the Black-Scholes model and appears in most financial asset models. Then we shall state the concept of martingale in a continuous-time set-up and, finally, we shall construct the stochastic integral and introduce the differential calculus associated with it, namely the Itô calculus.

It is advisable that, upon first reading, the reader passes over the proofs in small print, as they are very technical.

3.1 General comments on continuous-time processes

What do we exactly mean by *continuous-time processes*?

Definition 3.1.1 *A* continuous-time *stochastic process in a space E endowed with a σ- algebra $\mathcal{E}$ is a family $(X_t)_{t \in \mathbb{R}^+}$ of random variables defined on a probability space $(\Omega, \mathcal{A}, \mathbf{P})$ with values in a measurable space $(E, \mathcal{E})$.*

Remark 3.1.2

- In practice, the index t stands for the time.

- A process can also be considered as a random map: for each ω in Ω we associate the map from $\mathbb{R}^+$ to E: $t \to X_t(\omega)$, called a *path* of the process.

- A process can be considered as a map from $\mathbb{R}^+ \times \Omega$ into E. We shall always consider that this map is measurable when we endow the product set $\mathbb{R}^+ \times \Omega$ with the product σ-algebra $\mathcal{B}(\mathbb{R}^+) \times \mathcal{A}$ and when the set E is endowed with $\mathcal{E}$.

- We will only work with processes that are indexed on a finite time interval $[0, T]$.

As in discrete-time, we introduce the concept of *filtration*.

Definition 3.1.3 *Consider the probability space $(\Omega, \mathcal{A}, \mathbf{P})$, a* filtration $(\mathcal{F}_t)_{t \geq 0}$ *is an increasing family of σ-algebras included in $\mathcal{A}$.*

The σ-algebra $\mathcal{F}_t$ represents the information available at time t. We say that a process $(X_t)_{t \geq 0}$ is *adapted* to $(\mathcal{F}_t)_{t \geq 0}$, if for any t, X_t is $\mathcal{F}_t$-measurable.

Remark 3.1.4 From now on, we will be working with filtrations which have the following property

$$\text{If } A \in \mathcal{A} \text{ and if } \mathbf{P}(A) = 0, \text{ then for any } t, \ A \in \mathcal{F}_t.$$

In other words $\mathcal{F}_t$ contains all the $\mathbf{P}$-null sets of $\mathcal{A}$. The importance of this technical assumption is that if $X = Y$ $\mathbf{P}$ a.s. and Y is $\mathcal{F}_t$-measurable then we can show that X is also $\mathcal{F}_t$-measurable.

We can build a filtration generated by a process $(X_t)_{t \geq 0}$ and we write $\mathcal{F}_t = \sigma(X_s, s \leq t)$. In general, this filtration does not satisfy the previous condition. However, if we replace $\mathcal{F}_t$ by $\bar{\mathcal{F}}_t$ which is the σ-algebra generated by both $\mathcal{F}_t$ and $\mathcal{N}$ (the σ-algebra generated by all the $\mathbf{P}$-null sets of $\mathcal{A}$), we obtain a proper filtration satisfying the desired condition. We call it the *natural filtration* of the process $(X_t)_{t \geq 0}$. When we talk about a filtration without mentioning anything, it is assumed that we are dealing with the natural filtration of the process that we are considering. Obviously, a process is adapted to its natural filtration.

As in discrete-time, the concept of *stopping time* will be useful. A stopping time is a random time that depends on the underlying process in a non- anticipative way. In other words, at a given time t, we *know* if the stopping time is smaller than t. Formally, the definition is the following:

Definition 3.1.5 τ *is a* stopping time *with respect to the filtration $(\mathcal{F}_t)_{t \geq 0}$ if τ is a random variable in $\mathbb{R}^+ \cup \{+\infty\}$, such that for any $t \geq 0$*

$$\{\tau \leq t\} \in \mathcal{F}_t.$$

The σ-algebra associated with τ is defined as

$$\mathcal{F}_\tau = \{A \in \mathcal{A}, \text{ for any } t \geq 0, A \cap \{\tau \leq t\} \in \mathcal{F}_t\}.$$

This σ-algebra represents the information available before the random time τ. One can prove that (refer to Exercises 8, 9, 10, 11 and 14):

Proposition 3.1.6

- *If S is a stopping time, S is $\mathcal{F}_S$ measurable.*
- *If S is a stopping time, finite almost surely, and $(X_t)_{t\geq 0}$ is a continuous, adapted process, then X_S is $\mathcal{F}_S$ measurable.*
- *If S and T are two stopping times such that $S \leq T$ $\mathbf{P}$ a.s., then $\mathcal{F}_S \subset \mathcal{F}_T$.*
- *If S and T are two stopping times, then $S \wedge T = \inf(S, T)$ is a stopping time. In particular, if S is a stopping time and t is a deterministic time $S \wedge t$ is a stopping time.*

3.2 Brownian motion

A particularly important example of stochastic process is the *Brownian motion*. It will be the core of most financial models, whether we consider stocks, currencies or interest rates.

Definition 3.2.1 *A Brownian motion is a real-valued, continuous stochastic process $(X_t)_{t\geq 0}$, with independent and stationary increments. In other words:*

- *continuity: $\mathbf{P}$ a.s. the map $s \mapsto X_s(\omega)$ is continuous.*
- *independent increments: If $s \leq t$, $X_t - X_s$ is independent of $\mathcal{F}_s = \sigma(X_u, u \leq s)$.*
- *stationary increments: if $s \leq t$, $X_t - X_s$ and $X_{t-s} - X_0$ have the same probability law.*

This definition induces the distribution of the process X_t, but the result is difficult to prove and the reader ought to consult the book by Gihman and Skorohod (1980) for a proof of the following theorem.

Theorem 3.2.2 *If $(X_t)_{t\geq 0}$ is a Brownian motion, then $X_t - X_0$ is a normal random variable with mean rt and variance $\sigma^2 t$, where r and σ are constant real numbers.*

Remark 3.2.3 A Brownian motion is *standard* if

$$X_0 = 0 \ \mathbf{P} \text{ a.s.} \qquad \mathbf{E}(X_t) = 0, \qquad \mathbf{E}(X_t^2) = t.$$

From now on, a Brownian motion is assumed to be standard if nothing else is mentioned. In that case, the distribution of X_t is the following:

$$\frac{1}{\sqrt{2\pi t}} \exp\left(-\frac{x^2}{2t}\right) dx,$$

where dx is the Lebesgue measure on $\mathbb{R}$.

The following theorem emphasises the Gaussian property of the Brownian motion. We have just seen that for any t, X_t is a normal random variable. A stronger result is the following:

Theorem 3.2.4 *If* $(X_t)_{t\geq 0}$ *is a Brownian motion and if* $0 \leq t_1 < \cdots < t_n$ *then* $(X_{t_1}, \ldots, X_{t_n})$ *is a* Gaussian *vector.*

The reader ought to consult the Appendix, page 173, to recall some properties of Gaussian vectors.

Proof. Consider $0 \leq t_1 < \cdots < t_n$, then the random vector $(X_{t_1}, X_{t_2} - X_{t_1}, \ldots, X_{t_n} - X_{t_{n-1}})$ is composed of normal, independent random variables (by Theorem 3.2.2 and by definition of the Brownian motion). Therefore, this vector is Gaussian and so is $(X_{t_1}, \ldots, X_{t_n})$. $\square$

We shall also need a definition of a Brownian motion with respect to a filtration $(\mathcal{F}_t)$.

Definition 3.2.5 *A real-valued continuous stochastic process is an* $(\mathcal{F}_t)$-Brownian motion *if it satisfies:*

- *For any* $t \geq 0$, X_t *is* $\mathcal{F}_t$-measurable.
- *If* $s \leq t$, $X_t - X_s$ *is independent of the σ-algebra $\mathcal{F}_s$.*
- *If* $s \leq t$, $X_t - X_s$ *and* $X_{t-s} - X_0$ *have the same law.*

Remark 3.2.6 The first point of this definition shows that $\sigma(X_u, u \leq t) \subset \mathcal{F}_t$. Moreover, it is easy to check that an $\mathcal{F}_t$-Brownian motion is also a Brownian motion with respect to its natural filtration.

3.3 Continuous-time martingales

As in discrete-time models, the concept of martingale is a crucial tool to explain the notion of arbitrage. The following definition is an extension of the one in discrete-time.

Definition 3.3.1 *Let us consider a probability space* $(\Omega, \mathcal{A}, \mathbf{P})$ *and a filtration* $(\mathcal{F}_t)_{t\geq 0}$ *on this space. An adapted family* $(M_t)_{t\geq 0}$ *of integrable random variables, i.e.* $\mathbf{E}(|M_t|) < +\infty$ *for any t is:*

- a martingale *if, for any* $s \leq t$, $\mathbf{E}(M_t|\mathcal{F}_s) = M_s$;
- a supermartingale *if, for any* $s \leq t$, $\mathbf{E}(M_t|\mathcal{F}_s) \leq M_s$;
- a submartingale *if, for any* $s \leq t$, $\mathbf{E}(M_t|\mathcal{F}_s) \geq M_s$.

Remark 3.3.2 It follows from this definition that, if $(M_t)_{t\geq 0}$ is a martingale, then $\mathbf{E}(M_t) = \mathbf{E}(M_0)$ for any t.

Here are some examples of martingales.

Proposition 3.3.3 *If* $(X_t)_{t\geq 0}$ *is a standard $\mathcal{F}_t$-Brownian motion:*

1. X_t *is an* $\mathcal{F}_t$-martingale.
2. $X_t^2 - t$ *is an* $\mathcal{F}_t$-martingale.

3. $\exp\left(\sigma X_t - (\sigma^2/2)t\right)$ *is an* $\mathcal{F}_t$*-martingale.*

Proof. If $s \leq t$ then $X_t - X_s$ is independent of the σ-algebra $\mathcal{F}_s$. Thus $\mathbf{E}(X_t - X_s|\mathcal{F}_s) = \mathbf{E}(X_t - X_s)$. Since a standard Brownian motion has an expectation equal to zero, we have $\mathbf{E}(X_t - X_s) = 0$. Hence the first assertion is proved. To show the second one, we remark that

$$
\begin{aligned}
\mathbf{E}\left(X_t^2 - X_s^2|\mathcal{F}_s\right) &= \mathbf{E}\left((X_t - X_s)^2 + 2X_s(X_t - X_s)|\mathcal{F}_s\right) \\
&= \mathbf{E}\left((X_t - X_s)^2|\mathcal{F}_s\right) + 2X_s\mathbf{E}\left(X_t - X_s|\mathcal{F}_s\right),
\end{aligned}
$$

and since $(X_t)_{t\geq 0}$ is a martingale $\mathbf{E}\left(X_t - X_s|\mathcal{F}_s\right) = 0$, whence

$$
\mathbf{E}\left(X_t^2 - X_s^2|\mathcal{F}_s\right) = \mathbf{E}\left((X_t - X_s)^2|\mathcal{F}_s\right).
$$

Because the Brownian motion has independent and stationary increments, it follows that

$$
\begin{aligned}
\mathbf{E}\left((X_t - X_s)^2|\mathcal{F}_s\right) &= \mathbf{E}\left(X_{t-s}^2\right) \\
&= t - s.
\end{aligned}
$$

The last equality is due to the fact that X_t has a normal distribution with mean zero and variance t. That yields $\mathbf{E}\left(X_t^2 - t|\mathcal{F}_s\right) = X_s^2 - s$, if $s < t$.

Finally, let us recall that if g is a standard normal random variable, we know that

$$
\mathbf{E}\left(e^{\lambda g}\right) = \int_{-\infty}^{+\infty} e^{\lambda x} e^{-x^2/2} \frac{dx}{\sqrt{2\pi}} = e^{\lambda^2/2}.
$$

On the other hand, if $s < t$

$$
\mathbf{E}\left(e^{\sigma X_t - \sigma^2 t/2}|\mathcal{F}_s\right) = e^{\sigma X_s - \sigma^2 t/2}\mathbf{E}\left(e^{\sigma(X_t - X_s)}|\mathcal{F}_s\right)
$$

because X_s is $\mathcal{F}_s$-measurable. Since $X_t - X_s$ is independent of $\mathcal{F}_s$, it turns out that

$$
\begin{aligned}
\mathbf{E}\left(e^{\sigma(X_t - X_s)}|\mathcal{F}_s\right) &= \mathbf{E}\left(e^{\sigma(X_t - X_s)}\right) \\
&= \mathbf{E}\left(e^{\sigma X_{t-s}}\right) \\
&= \mathbf{E}\left(e^{\sigma g\sqrt{t-s}}\right) \\
&= \exp\left(\frac{1}{2}\sigma^2(t - s)\right)
\end{aligned}
$$

That completes the proof. $\qquad\qquad\square$

If $(M_t)_{t\geq 0}$ is a martingale, the property $\mathbf{E}\left(M_t|\mathcal{F}_s\right) = M_s$, is also true if t and s are *bounded stopping times*. This result is actually an adaptation of Exercise 1 in Chapter 2 to the continuous case and it is called the *optional sampling theorem*. We will not prove this theorem, but the reader ought to refer to Karatzas and Shreve (1988), page 19.

Theorem 3.3.4 (optional sampling theorem) *If $(M_t)_{t\geq 0}$ is a continuous martingale with respect to the filtration $(\mathcal{F}_t)_{t\geq 0}$, and if τ_1 and τ_2 are two stopping times such that $\tau_1 \leq \tau_2 \leq K$, where K is a finite real number, then M_{τ_2} is integrable and*

$$\mathbf{E}\left(M_{\tau_2}|\mathcal{F}_{\tau_1}\right) = M_{\tau_1} \quad \mathbf{P} \ a.s.$$

Remark 3.3.5

- This result implies that if τ is a bounded stopping time then $\mathbf{E}(M_\tau) = \mathbf{E}(M_0)$ (apply the theorem with $\tau_1 = 0$, $\tau_2 = \tau$ and take the expectation on both sides).

- If M_t is a submartingale, the same theorem is true if we replace the previous equality by

$$\mathbf{E}\left(M_{\tau_2}|\mathcal{F}_{\tau_1}\right) \geq M_{\tau_1} \quad \mathbf{P} \ a.s.$$

We shall now apply that result to study the properties of the *hitting time* of a point by a Brownian motion.

Proposition 3.3.6 *Once again, we consider $(X_t)_{t\geq 0}$ an $\mathcal{F}_t$-Brownian motion. If a is a real number, we call $T_a = \inf\{s \geq 0, X_s = a\}$ or $+\infty$ if that set is empty.*

Then, T_a is a stopping time, finite almost surely, and its distribution is characterised by its Laplace transform

$$\mathbf{E}\left(e^{-\lambda T_a}\right) = e^{-\sqrt{2\lambda}|a|}.$$

Proof. We will assume that $a \geq 0$. First, we show that T_a is a stopping time. Indeed, since X_s is continuous

$$\{T_a \leq t\} = \cap_{\epsilon\in\mathbf{Q}^{+*}}\left\{\sup_{s\leq t} X_s > a - \epsilon\right\} = \cap_{\epsilon\in\mathbf{Q}^{+*}}\cup_{s\in\mathbf{Q}^+, s\leq t}\{X_s > a - \epsilon\}.$$

That last set belongs to $\mathcal{F}_t$, and therefore the result is proved. In the following, we write $x \wedge y = \inf(x, y)$.

Let us apply the sampling theorem to the martingale $M_t = \exp\left(\sigma X_t - (\sigma^2/2)t\right)$. We cannot apply the theorem to T_a which is not necessarily bounded; however, if n is a positive integer, $T_a \wedge n$ is a bounded stopping time (see Proposition 3.1.6), and from the optional sampling theorem

$$\mathbf{E}\left(M_{T_a \wedge n}\right) = 1.$$

On the one hand $M_{T_a \wedge n} = \exp\left(\sigma X_{T_a \wedge n} - \sigma^2(T_a \wedge n)/2\right) \leq \exp\left(\sigma a\right)$. On the other hand, if $T_a < +\infty$, $\lim_{n\to+\infty} M_{T_a \wedge n} = M_{T_a}$ and if $T_a = +\infty$, $X_t \leq a$ at any t, therefore $\lim_{n\to+\infty} M_{T_a \wedge n} = 0$. Lebesgue theorem implies that $\mathbf{E}(1_{\{T_a<+\infty\}}M_{T_a}) = 1$, i.e. since $X_{T_a} = a$ when $T_a < +\infty$

$$\mathbf{E}\left(1_{\{T_a<+\infty\}} \exp\left(-\frac{\sigma^2}{2}T_a\right)\right) = e^{-\sigma a}.$$

By letting σ converge to 0, we show that $\mathbf{P}(T_a < +\infty) = 1$ (which means that the Brownian motion reaches the level a almost surely). Also

$$\mathbf{E}\left(\exp\left(-\frac{\sigma^2}{2}T_a\right)\right) = e^{-\sigma a}.$$

The case $a < 0$ is easily solved if we notice that

$$T_a = \inf\{s \geq 0, -X_s = -a\},$$

where $(-X_t)_{t\geq 0}$ is an $\mathcal{F}_t$-Brownian motion because it is a continuous stochastic process with zero mean and variance t and with stationary, independent increments. □

The optional sampling theorem is also very useful to compute expectations involving the running maximum of a martingale. If M_t is a square integrable martingale, we can show that the second-order moment of $\sup_{0\leq t\leq T}|M_t|$ can be bounded. This is known as the *Doob inequality*.

Theorem 3.3.7 (Doob inequality) *If* $(M_t)_{0\leq t\leq T}$ *is a continuous martingale, we have*

$$\mathbf{E}\left(\sup_{0\leq t\leq T}|M_t|^2\right) \leq 4\mathbf{E}(|M_T|^2).$$

The proof of this theorem is the purpose of Exercise 13.

3.4 Stochastic integral and Itô calculus

In a discrete-time model, if we follow a self-financing strategy $\phi = (H_n)_{0\leq n\leq N}$, the discounted value of the portfolio with initial wealth V_0 is

$$V_0 + \sum_{j=1}^{n} H_j(\tilde{S}_j - \tilde{S}_{j-1}).$$

That wealth appears to be a *martingale transform* under a certain probability measure such that the discounted price of the stock is a martingale. As far as continuous-time models are concerned, integrals of the form $\int H_s d\tilde{S}_s$ will help us to describe the same idea.

However, the processes modelling stock prices are normally functions of one or several Brownian motions. But one of the most important properties of a Brownian motion is that, almost surely, its paths are not differentiable at any point. In other words, if (X_t) is a Brownian motion, it can be proved that for almost every $\omega \in \Omega$, there is not any time t in $\mathbb{R}^+$ such that dX_t/dt exists. As a result, we are not able to define the integral above as

$$\int_0^t f(s)dX_s = \int_0^t f(s)\frac{dX_s}{ds}ds.$$

Nevertheless, we are able to define this type of integral with respect to a Brownian motion, and we shall call them *stochastic integrals*. That is the whole purpose of this section.

3.4.1 Construction of the stochastic integral

Suppose that $(W_t)_{t\geq 0}$ is a standard $\mathcal{F}_t$-Brownian motion defined on a filtered probability space $(\Omega, \mathcal{A}, (\mathcal{F}_t)_{t\geq 0}, \mathbf{P})$. We are about to give a meaning to the expression $\int_0^t f(s, \omega)dW_s$ for a certain class of processes $f(s, \omega)$ adapted to the filtration $(\mathcal{F}_t)_{t\geq 0}$. To start with, we shall construct this stochastic integral for a set of processes called *simple processes*. Throughout the text, T will be a strictly positive, finite real number.

Definition 3.4.1 $(H_t)_{0\leq t\leq T}$ *is called a* simple process *if it can be written as*

$$H_t(\omega) = \sum_{i=1}^p \phi_i(\omega)\mathbf{1}_{]t_{i-1}, t_i]}(t)$$

where $0 = t_0 < t_1 < \cdots < t_p = T$ *and* ϕ_i *is* $\mathcal{F}_{t_{i-1}}$-*measurable and bounded.*

Then, by definition, the stochastic integral of a simple process H is the continuous process $(I(H)_t)_{0\leq t\leq T}$ defined for any $t \in]t_k, t_{k+1}]$ as

$$I(H)_t = \sum_{1\leq i\leq k} \phi_i(W_{t_i} - W_{t_{i-1}}) + \phi_{k+1}(W_t - W_{t_k}).$$

Note that $I(H)_t$ can be written as

$$I(H)_t = \sum_{1\leq i\leq p} \phi_i(W_{t_i\wedge t} - W_{t_{i-1}\wedge t}),$$

which proves the continuity of $t \mapsto I(H)_t$. We shall write $\int_0^t H_s dW_s$ for $I(H)_t$. The following proposition is fundamental.

Proposition 3.4.2 *If* $(H_t)_{0\leq t\leq T}$ *is a simple process:*

- $\left(\int_0^t H_s dW_s\right)_{0\leq t\leq T}$ *is a continuous* $\mathcal{F}_t$-*martingale,*

- $\mathbf{E}\left(\left(\int_0^t H_s dW_s\right)^2\right) = \mathbf{E}\left(\int_0^t H_s^2 ds\right),$

- $\mathbf{E}\left(\sup_{t\leq T}\left|\int_0^t H_s dW_s\right|^2\right) \leq 4\mathbf{E}\left(\int_0^T H_s^2 ds\right).$

Proof.   In order to prove this proposition, we are going to use discrete-time processes. Indeed, to show that $\left(\int_0^t H_s dW_s\right)$ is a martingale, we just need to check that, for any $t > s$,

$$\mathbf{E}\left(\int_0^t H_u dW_u|\mathcal{F}_s\right) = \int_0^s H_u dW_u.$$

If we include s and t to the subdivision $t_0 = 0 < t_1 < \cdots < t_p = T$, and if we call $M_n = \int_0^{t_n} H_s dW_s$ and $\mathcal{G}_n = \mathcal{F}_{t_n}$ for $0 \leq n \leq p$, we want to show that M_n is a $\mathcal{G}_n$-martingale. To prove it, we notice that

$$M_n = \int_0^{t_n} H_s dW_s = \sum_{i=1}^{n} \phi_i \left(W_{t_i} - W_{t_{i-i}} \right)$$

with ϕ_i $\mathcal{G}_{i-1}$-measurable. Moreover, $X_n = W_{t_n}$ is a $\mathcal{G}_n$-martingale (since $(W_t)_{t \geq 0}$ is a Brownian motion). $(M_n)_{n \in [0,p]}$ turns out to be a martingale transform of $(X_n)_{n \in [0,p]}$. The Proposition 1.2.3 of Chapter 1 allows us to conclude that $(M_n)_{n \in [0,p]}$ is a martingale. The second assertion comes from the fact that

$$
\begin{aligned}
\mathbf{E}(M_n^2) &= \mathbf{E} \left(\left(\sum_{i=1}^{n} \phi_i (X_i - X_{i-1}) \right)^2 \right) \\
&= \sum_{i=1}^{n} \sum_{j=1}^{n} \mathbf{E} \left(\phi_i \phi_j (X_i - X_{i-1})(X_j - X_{j-1}) \right). \quad (3.1)
\end{aligned}
$$

Also, if $i < j$, we have

$$
\begin{aligned}
&\mathbf{E} \left(\phi_i \phi_j (X_i - X_{i-1})(X_j - X_{j-1}) \right) \\
&= \mathbf{E} \left(\mathbf{E} \left(\phi_i \phi_j (X_i - X_{i-1})(X_j - X_{j-1}) | \mathcal{G}_{j-1} \right) \right) \\
&= \mathbf{E} \left(\phi_i \phi_j (X_i - X_{i-1}) \mathbf{E} \left(X_j - X_{j-1} | \mathcal{G}_{j-1} \right) \right).
\end{aligned}
$$

Since X_j is a martingale, $\mathbf{E}(X_j - X_{j-1} | \mathcal{G}_{j-1}) = 0$. Therefore, if $i < j$, $\mathbf{E} \left(\phi_i \phi_j (X_i - X_{i-1})(X_j - X_{j-1}) \right) = 0$. If $j > i$ we get the same thing. Finally, if $i = j$,

$$
\begin{aligned}
\mathbf{E} \left(\phi_i^2 (X_i - X_{i-1})^2 \right) &= \mathbf{E} \left(\mathbf{E} \left(\phi_i^2 (X_i - X_{i-1})^2 | \mathcal{G}_{i-1} \right) \right) \\
&= \mathbf{E} \left(\phi_i^2 \mathbf{E} \left((X_i - X_{i-1})^2 | \mathcal{G}_{i-1} \right) \right),
\end{aligned}
$$

as a result

$$\mathbf{E} \left((X_i - X_{i-1})^2 | \mathcal{G}_{i-1} \right) = E \left((W_{t_i} - W_{t_{i-1}})^2 \right) = t_i - t_{i-1}. \quad (3.2)$$

From (3.1) and (3.2) we conclude that

$$\mathbf{E} \left(\left(\sum_{i=1}^{n} \phi_i (X_i - X_{i-1}) \right)^2 \right) = \mathbf{E} \left(\sum_{i=1}^{n} \phi_i^2 (t_i - t_{i-1}) \right) = \mathbf{E} \left(\int_0^t H_s^2 ds \right).$$

The continuity of $t \rightarrow \int_0^t H_s dW_s$ is clear if we look at the definition. The third assertion is just a consequence of Doob inequality (3.3.7) applied to the continuous martingale $\left(\int_0^t H_s dW_s \right)_{t \geq 0}$. $\qquad \square$

Remark 3.4.3 We write by definition

$$\int_t^T H_s dW_s = \int_0^T H_s dW_s - \int_0^t H_s dW_s.$$

If $t \leq T$, and if $A \in \mathcal{F}_t$, then $s \to 1_A 1_{\{t<s\}} H_s$ is still a simple process and it is easy to check from the definition of the integral that

$$\int_0^T 1_A H_s 1_{\{t<s\}} dW_s = 1_A \int_t^T H_s dW_s. \qquad (3.3)$$

Now that we have defined the stochastic integral for simple processes and stated some of its properties, we are going to extend the concept to a larger class of adapted processes $\mathcal{H}$

$$\mathcal{H} = \left\{ (H_t)_{0 \leq t \leq T}, (\mathcal{F}_t)_{t \geq 0} - \text{adapted process, } \mathbf{E} \left(\int_0^T H_s^2 ds \right) < +\infty \right\}.$$

Proposition 3.4.4 *Consider* $(W_t)_{t \geq 0}$ *an* $\mathcal{F}_t$-*Brownian motion. There exists a unique linear mapping* J *from* $\mathcal{H}$ *to the space of continuous* $\mathcal{F}_t$-*martingales defined on* $[0, T]$, *such that:*

1. If $(H_t)_{t \leq T}$ *is a simple process,* $\mathbf{P}$ *a.s. for any* $0 \leq t \leq T$, $J(H)_t = I(H)_t$.

2. If $t \leq T$, $\mathbf{E} \left(J(H)_t^2 \right) = \mathbf{E} \left(\int_0^t H_s^2 ds \right)$.

This linear mapping is unique in the following sense, if both J *and* J' *satisfy the previous properties then*

$$\mathbf{P} \text{ a.s.} \quad \forall 0 \leq t \leq T, \ J(H)_t = J'(H)_t.$$

We denote, for $H \in \mathcal{H}$, $\int_0^t H_s dW_s = J(H)_t$.

On top of that, the stochastic integral satisfies the following properties:

Proposition 3.4.5 *If* $(H_t)_{0 \leq t \leq T}$ *belongs to* $\mathcal{H}$ *then*

1. We have

$$\mathbf{E} \left(\sup_{t \leq T} \left| \int_0^t H_s dW_s \right|^2 \right) \leq 4 \mathbf{E} \left(\int_0^T H_s^2 ds \right). \qquad (3.4)$$

2. If τ *is an* $\mathcal{F}_t$-*stopping time*

$$\mathbf{P} \text{ a.s.} \quad \int_0^\tau H_s dW_s = \int_0^T 1_{\{s \leq \tau\}} H_s dW_s. \qquad (3.5)$$

Proof. We shall use the fact that if $(H_s)_{s \leq T}$ is in $\mathcal{H}$, there exists a sequence $(H_s^n)_{s \leq T}$ of simple processes such that

$$\lim_{n \to +\infty} \mathbf{E} \left(\int_0^T |H_s - H_s^n|^2 ds \right) = 0.$$

A proof of this result can be found in Karatzas and Shreve (1988) (page 134, problem 2.5).

If $H \in \mathcal{H}$ and $(H^n)_{n \geq 0}$ is a sequence of simple processes converging to H in the previous sense, we have

$$\mathbf{E}\left(\sup_{t \leq T} |I(H^{n+p})_t - I(H^n)_t|^2\right) \leq 4\mathbf{E}\left(\int_0^T |H_s^{n+p} - H_s^n|^2 \, ds\right). \quad (3.6)$$

Therefore, there exists a subsequence $H^{\phi(n)}$ such that

$$\mathbf{E}\left(\sup_{t \leq T} |I(H^{\phi(n+1)})_t - I(H^{\phi(n)})_t|^2\right) \leq \frac{1}{2^n}.$$

thus, the series whose general term is $I(H^{\phi(n+1)}) - I(H^{\phi(n)})$ is uniformly convergent, almost surely. Consequently $I(H^{\phi(n)})_t$ converges towards a continuous function which will be, by definition, the map $t \mapsto J(H)_t$. Taking the limit in (3.6), we obtain

$$\mathbf{E}\left(\sup_{t \leq T} |J(H)_t - I(H^n)_t|^2\right) \leq 4\mathbf{E}\left(\int_0^T |H_s - H_s^n|^2 \, ds\right). \quad (3.7)$$

That implies that $(J(H)_t)_{0 \leq t \leq T}$ does not depend on the approximating sequence. On the other hand, $(J(H)_t)_{0 \leq t \leq T}$ is a martingale, indeed

$$\mathbf{E}\left(I(H^n)_t | \mathcal{F}_s\right) = I(H^n)_s.$$

Moreover, for any t, $\lim_{n \to +\infty} I(H^n)_t = J(H)_t$ in $L^2(\Omega, \mathbf{P})$ norm and, because the conditional expectation is continuous in $L^2(\Omega, \mathbf{P})$, we can conclude.

From (3.7) and from the fact that $\mathbf{E}(I(H^n)_t^2) = \mathbf{E}\left(\int_0^T |H_s^n|^2 \, ds\right)$, it follows that $\mathbf{E}(J(H)_t^2) = \mathbf{E}\left(\int_0^T |H_s|^2 \, ds\right)$. In the same way, from (3.7) and since $\mathbf{E}(\sup_{t \leq T} I(H^n)_t^2) \leq 4\mathbf{E}\left(\int_0^T |H_s^n|^2 \, ds\right)$, we prove (3.4).

The uniqueness of the extension results from the fact that the set of simple processes is dense in $\mathcal{H}$.

We now prove (3.5). First of all, we notice that (3.3) is still true if $H \in \mathcal{H}$. This is justified by the fact that the simple processes are dense in $\mathcal{H}$ and by (3.7). We first consider stopping times of the form $\tau = \sum_{1 \leq i \leq n} t_i \mathbf{1}_{A_i}$, where $0 < t_1 < \cdots < t_n = T$ and the A_i's are disjoint and $\mathcal{F}_{t_i}$ measurable, and we prove (3.5) in that case. First

$$\int_0^T \mathbf{1}_{\{s > \tau\}} H_s \, dW_s = \int_0^T \left(\sum_{1 \leq i \leq n} \mathbf{1}_{A_i} \mathbf{1}_{\{s > t_i\}}\right) H_s \, dW_s,$$

also, each $1_{\{s>t_i\}}1_{A_i}H_s$ is adapted because this process is zero if $s \leq t_i$ and is equal to $1_{A_i}H_s$ otherwise, therefore it belongs to $\mathcal{H}$. It follows that

$$\int_0^T 1_{\{s>\tau\}}H_s dW_s = \sum_{1\leq i\leq n} \int_0^T 1_{A_i}1_{\{s>t_i\}}H_s dW_s$$

$$= \sum_{1\leq i\leq n} 1_{A_i}\int_{t_i}^T H_s dW_s = \int_\tau^T H_s dW_s,$$

and then $\int_0^T 1_{\{s\leq\tau\}}H_s dW_s = \int_0^\tau H_s dW_s$.

In order to prove this result for an arbitrary stopping time τ, we must notice that τ can be approximated by a decreasing sequence of stopping times of the previous form. If

$$\tau_n = \sum_{0\leq i\leq 2^n} \frac{(k+1)T}{2^n}1_{\left\{\frac{kT}{2^n}\leq\tau<\frac{(k+1)T}{2^n}\right\}}.$$

τ_n converges almost surely to τ. By continuity of the map $t \mapsto \int_0^t H_s dW_s$ we can affirm that, almost surely, $\int_0^{\tau_n} H_s dW_s$ converges to $\int_0^\tau H_s dW_s$. On the other hand

$$\mathbf{E}\left(\left|\int_0^T 1_{\{s\leq\tau\}}H_s dW_s - \int_0^T 1_{\{s\leq\tau_n\}}H_s dW_s\right|^2\right) = \mathbf{E}\left(\int_0^T 1_{\{\tau<s\leq\tau_n\}}H_s^2 ds\right).$$

This last term converges to 0 by dominated convergence, therefore $\int_0^T 1_{\{s\leq\tau_n\}}H_s dW_s$ converges to $\int_0^T 1_{\{s\leq\tau\}}H_s dW_s$ in $L^2(\Omega,\mathbf{P})$ (a subsequence converges almost surely). That completes the proof of (3.5) for an arbitrary stopping time. $\square$

In the modelling, we shall need processes which only satisfy a weaker integrability condition than the processes in $\mathcal{H}$, that is why we define

$$\tilde{\mathcal{H}} = \left\{(H_s)_{0\leq s\leq T} \ (\mathcal{F}_t)_{t\geq 0} - \text{adapted process}, \int_0^T H_s^2 ds < +\infty \ \mathbf{P} \text{ a.s.}\right\}.$$

The following proposition defines an extension of the stochastic integral from $\mathcal{H}$ to $\tilde{\mathcal{H}}$.

Proposition 3.4.6 *There exists a unique linear mapping $\tilde{J}$ from $\tilde{\mathcal{H}}$ into the vector space of* continuous processes *defined on $[0,T]$, such that:*

1. Extension property: *If $(H_t)_{0\leq t\leq T}$ is a simple process then*

$$\mathbf{P} \text{ a.s. } \forall 0 \leq t \leq T, \ \tilde{J}(H)_t = I(H)_t.$$

2. Continuity property: *If $(H^n)_{n\geq 0}$ is a sequence of processes in $\tilde{\mathcal{H}}$ such that $\int_0^T (H_s^n)^2 ds$ converges to 0 in probability then $\sup_{t\leq T} |\tilde{J}(H^n)_t|$ converges to 0 in probability.*

Consistently, we write $\int_0^t H_s dW_s$ for $\tilde{J}(H)_t$.

Remark 3.4.7 It is crucial to notice that in this case $\left(\int_0^t H_s dW_s\right)_{0 \leq t \leq T}$ is not *necessarily a martingale.*

Proof. It is easy to deduce from the extension property and from the continuity property that if $H \in \mathcal{H}$ then **P** a.s. $\forall t \leq T$, $\tilde{J}(H)_t = J(H)_t$.

Let $H \in \tilde{\mathcal{H}}$, and define $T_n = \inf\left\{0 \leq s \leq T, \int_0^s H_u^2 du \geq n\right\}$ ($+\infty$ if that set is empty), and $H_s^n = H_s 1_{\{s \leq T_n\}}$.

Firstly, we show that T_n is a stopping time. Since $\{T_n \leq t\} = \{\int_0^t H_u^2 du \geq n\}$, we just need to prove that $\int_0^t H_u^2 du$ is $\mathcal{F}_t$-measurable. This result is true if H is a simple process and, by density, it is true if $H \in \mathcal{H}$. Finally, if $H \in \tilde{\mathcal{H}}$, $\int_0^t H_u^2 du$ is also $\mathcal{F}_t$-measurable because it is the limit of $\int_0^t (H_u \wedge K)^2 du$ almost surely as K tends to infinity. Then, we see immediately that the processes H_s^n are adapted and bounded, hence they belong to $\mathcal{H}$. Moreover

$$\int_0^t H_s^n dW_s = \int_0^t 1_{\{s \leq T_n\}} H_s^{n+1} dW_s,$$

and relation (3.5) implies that

$$\int_0^t H_s^n dW_s = \int_0^{t \wedge T_n} H_s^{n+1} dW_s.$$

Thus, on the set $\{\int_0^T H_u^2 du < n\}$, for any $t \leq T$, $J(H^n)_t = J(H^{n+1})_t$. Since $\cup_{n \geq 0}\{\int_0^T H_u^2 du < n\} = \{\int_0^T H_u^2 du < +\infty\}$, we can define almost surely a process $\tilde{J}(H)_t$ by: on the set $\{\int_0^T H_u^2 du < n\}$,

$$\forall t \leq T \quad \tilde{J}(H)_t = J(H^n)_t.$$

The process $t \mapsto \tilde{J}(H)_t$ is almost surely continuous, by definition. The extension property is satisfied by construction. We just need to prove the continuity property of $\tilde{J}$. To do so, we first notice that

$$\mathbf{P}\left(\sup_{t \leq T}\left|\tilde{J}(H)_t\right| \geq \epsilon\right) \leq \mathbf{P}\left(\int_0^T H_s^2 ds \geq 1/N\right)$$
$$+ \mathbf{P}\left(1_{\{\int_0^T H_u^2 du < 1/N\}} \sup_{t \leq T}\left|\tilde{J}(H)_t\right| \geq \epsilon\right).$$

If we call $\tau_N = \inf\left\{s \leq T, \int_0^s H_u^2 du \geq 1/N\right\}$ ($+\infty$ if this set is empty), then on the set $\left\{\int_0^T H_u^2 du < 1/N\right\}$, it follows from (3.5) that, for any $t \leq T$,

$$\int_0^t H_s dW_s = \tilde{J}(H)_t = J(H^1)_t = \int_0^t H_s^1 1_{\{s \leq \tau_N\}} dW_s = \int_0^t H_s 1_{\{s \leq \tau_N\}} dW_s,$$

whence, by applying (3.4) to the process $s \mapsto H_s 1_{\{s \leq \tau_N\}}$ we get

$$
\mathbf{P}\left(\sup_{t \leq T}\left|\tilde{J}(H)_t\right| \geq \epsilon\right) \;\leq\; \mathbf{P}\left(\int_0^T H_s^2 ds \geq \frac{1}{N}\right)
$$
$$
+4/\epsilon^2 \mathbf{E}\left(\int_0^T H_s^2 1_{\{s \leq \tau_N\}} ds\right)
$$
$$
\leq\; \mathbf{P}\left(\int_0^T H_s^2 ds \geq \frac{1}{N}\right) + \frac{4}{N\epsilon^2}.
$$

As a result, if $\int_0^T (H_s^n)^2 ds$ converges to 0 in probability, then $\sup_{t \leq T}|\tilde{J}(H^n)_t|$ converges to 0 in probability.

In order to prove the linearity of $\tilde{J}$, let us consider two processes belonging to $\tilde{\mathcal{H}}$, called H and K, and the two sequences H_t^n and K_t^n defined at the beginning of the proof such that $\int_0^T (H_s^n - H_s)^2 ds$ and $\int_0^T (K_s^n - K_s)^2 ds$ converge to 0 in probability. By continuity of $\tilde{J}$ we can take the limit in the equality $J(\lambda H^n + \mu K^n)_t = \lambda J(H^n)_t + \mu J(K^n)_t$, to prove the continuity of $\tilde{J}$.

Finally, the fact that if $H \in \tilde{\mathcal{H}}$ then $\int_0^T (H_t - H_t^n)^2 dt$ converge to 0 in probability and the continuity property yields the uniqueness of the extension.   □

We are about to summarise the conditions needed to define the stochastic integral with respect to a Brownian motion and we want to specify the assumptions that make it a martingale.

Summary:

Let us consider an $\mathcal{F}_t$-Brownian motion $(W_t)_{t \geq 0}$ and an $\mathcal{F}_t$-adapted process $(H_t)_{0 \leq t \leq T}$. We are able to define the stochastic integral $(\int_0^t H_s dW_s)_{0 \leq t \leq T}$ as soon as $\int_0^T H_s^2 ds < +\infty$ **P** a.s. By construction, the process $(\int_0^t H_s dW_s)_{0 \leq t \leq T}$ is a martingale *if* $\mathbf{E}\left(\int_0^T H_s^2 ds\right) < +\infty$. This condition is not necessary. Indeed, the inequality $\mathbf{E}\left(\int_0^T H_s^2 ds\right) < +\infty$ is satisfied if and only if

$$
\mathbf{E}\left(\sup_{0 \leq t \leq T}\left(\int_0^t H_s dW_s\right)^2\right) < +\infty.
$$

This is proved in Exercise 15.

3.4.2 Itô calculus

It is now time to introduce a differential calculus based on this stochastic integral. It will be called the *Itô calculus* and the main ingredient is the famous *Itô formula*.

In particular, the Itô formula allows us to differentiate such a function as $t \mapsto f(W_t)$ if f is twice continuously differentiable. The following example will simply show that a naive extension of the classical differential calculus is bound to fail. Let us try to *differentiate* the function $t \to W_t^2$ in terms of 'dW_t'. Typically, for a differentiable function $f(t)$ null at the origin, we have $f(t)^2 = 2\int_0^t f(s)\dot{f}(s)ds = 2\int_0^t f(s)df(s)$. In the Brownian case, it is impossible to have a similar formula $W_t^2 = 2\int_0^t W_s dW_s$. Indeed, from the previous section we know that $\int_0^t W_s dW_s$ is a martingale (because $\mathbf{E}\left(\int_0^t W_s^2 ds\right) < +\infty$), null at zero. If it were equal to $W_t^2/2$ it would be non-negative, and a non-negative martingale vanishing at zero can only be identically zero.

We shall define precisely the class of processes for which the Itô formula is applicable.

Definition 3.4.8 *Let* $(\Omega, \mathcal{F}, (\mathcal{F}_t)_{t\geq 0}, \mathbf{P})$ *be a filtered probability space and* $(W_t)_{t\geq 0}$ *an* $\mathcal{F}_t$-*Brownian motion.* $(X_t)_{0\leq t\leq T}$ *is an* $\mathbb{R}$-*valued* Itô *process if it can be written as*

$$\mathbf{P} \text{ a.s. } \forall t \leq T \quad X_t = X_0 + \int_0^t K_s ds + \int_0^t H_s dW_s,$$

where

- X_0 *is* $\mathcal{F}_0$-*measurable.*
- $(K_t)_{0\leq t\leq T}$ *and* $(H_t)_{0\leq t\leq T}$ *are* $\mathcal{F}_t$-*adapted processes.*
- $\int_0^T |K_s|ds < +\infty$ $\mathbf{P}$ *a.s.*
- $\int_0^T |H_s|^2 ds < +\infty$ $\mathbf{P}$ *a.s.*

We can prove the following proposition (see Exercise 16) which underlines the uniqueness of the previous decomposition.

Proposition 3.4.9 *If* $(M_t)_{0\leq t\leq T}$ *is a continuous martingale such that*

$$M_t = \int_0^t K_s ds, \text{ with } \mathbf{P} \text{ a.s. } \int_0^T |K_s|ds < +\infty,$$

then

$$\mathbf{P} \text{ a.s. } \forall t \leq T, \ M_t = 0.$$

This implies that:

– *An Itô process decomposition is unique. That means that if*

$$X_t = X_0 + \int_0^t K_s ds + \int_0^t H_s dW_s = X_0' + \int_0^t K_s' ds + \int_0^t H_s' dW_s$$

then

$$X_0 = X_0' \quad d\mathbf{P} \text{ a.s.} \quad H_s = H_s' \quad ds \times d\mathbf{P} \text{ a.e.} \quad K_s = K_s' \quad ds \times d\mathbf{P} \text{ a.e.}$$

– *If* $(X_t)_{0\leq t\leq T}$ *is a martingale of the form* $X_0 + \int_0^t K_s ds + \int_0^t H_s dW_s$, *then* $K_t = 0$ $dt \times d\mathbf{P}$ *a.e.*

We shall state Itô formula for continuous martingales. The interested reader should refer to Bouleau (1988) for an elementary proof in the Brownian case, i.e. when (W_t) is a standard Brownian motion, or to Karatzas and Shreve (1988) for a complete proof.

Theorem 3.4.10 *Let* $(X_t)_{0 \le t \le T}$ *be an Itô process*

$$X_t = X_0 + \int_0^t K_s ds + \int_0^t H_s dW_s,$$

and f *be a twice continuously differentiable function, then*

$$f(X_t) = f(X_0) + \int_0^t f'(X_s)dX_s + \frac{1}{2}\int_0^t f''(X_s)d\langle X, X\rangle_s$$

where, by definition

$$\langle X, X\rangle_t = \int_0^t H_s^2 ds,$$

and

$$\int_0^t f'(X_s)dX_s = \int_0^t f'(X_s)K_s ds + \int_0^t f'(X_s)H_s dW_s.$$

Likewise, if $(t, x) \to f(t, x)$ *is a function which is twice differentiable with respect to* x *and once with respect to* t, *and if these partial derivatives are continuous with respect to* (t, x) *(i.e.* f *is a function of class* $C^{1,2}$), *Itô formula becomes*

$$
\begin{aligned}
f(t, X_t) \;=\; & f(0, X_0) + \int_0^t f_s'(s, X_s)ds \\
& + \int_0^t f_x'(s, X_s)dX_s + \frac{1}{2}\int_0^t f_{xx}''(s, X_s)d\langle X, X\rangle_s.
\end{aligned}
$$

3.4.3 Examples: Itô formula in practice

Let us start by giving an elementary example. If $f(x) = x^2$ and $X_t = W_t$, we identify $K_s = 0$ and $H_s = 1$, thus

$$W_t^2 = 2\int_0^t W_s dW_s + \frac{1}{2}\int_0^t 2ds.$$

It turns out that

$$W_t^2 - t = 2\int_0^t W_s dW_s.$$

Since $\mathbf{E}\left(\int_0^t W_s^2 ds\right) < +\infty$, it confirms the fact that $W_t^2 - t$ is a martingale.

We now want to tackle the problem of finding the solutions $(S_t)_{t \ge 0}$ of

$$S_t = x_0 + \int_0^t S_s\left(\mu ds + \sigma dW_s\right). \tag{3.8}$$

This is often written in the symbolic form

$$dS_t = S_t\left(\mu dt + \sigma dW_t\right), \quad S_0 = x_0. \tag{3.9}$$

We are actually looking for an adapted process $(S_t)_{t \ge 0}$ such that the integrals

$\int_0^t S_s ds$ and $\int_0^t S_s dW_s$ exist and at any time t

$$\mathbf{P} \text{ a.s. } S_t = x_0 + \int_0^t \mu S_s ds + \int_0^t \sigma S_s dW_s.$$

To put it in a simple way, let us do a formal calculation. We write $Y_t = \log(S_t)$ where S_t is a solution of (3.8). S_t is an Itô process with $K_s = \mu S_s$ and $H_s = \sigma S_s$. Assuming that S_t is non-negative, we apply Itô formula to $f(x) = \log(x)$ (at least formally, because $f(x)$ is not a C^2 function!), and we obtain

$$\log(S_t) = \log(S_0) + \int_0^t \frac{dS_s}{S_s} + \frac{1}{2} \int_0^t \left(\frac{-1}{S_s^2} \right) \sigma^2 S_s^2 ds.$$

Using (3.9), we get

$$Y_t = Y_0 + \int_0^t \left(\mu - \sigma^2/2 \right) dt + \int_0^t \sigma dW_t,$$

and finally

$$Y_t = \log(S_t) = \log(S_0) + \left(\mu - \sigma^2/2 \right) t + \sigma W_t.$$

Taking that into account, it seems that

$$S_t = x_0 \exp \left(\left(\mu - \sigma^2/2 \right) t + \sigma W_t \right)$$

is a solution of equation (3.8). We must check that conjecture rigorously. We have $S_t = f(t, W_t)$ with

$$f(t, x) = x_0 \exp \left(\left(\mu - \sigma^2/2 \right) t + \sigma x \right).$$

Itô formula is now applicable and yields

$$
\begin{aligned}
S_t &= f(t, W_t) \\
&= f(0, W_0) + \int_0^t f_s'(s, W_s) ds \\
&\quad + \int_0^t f_x'(s, W_s) dW_s + \frac{1}{2} \int_0^t f_{xx}''(s, W_s) d\langle W, W \rangle_s.
\end{aligned}
$$

Furthermore, since $\langle W, W \rangle_t = t$, we can write

$$S_t = x_0 + \int_0^t S_s \left(\mu - \sigma^2/2 \right) ds + \int_0^t S_s \sigma dW_s + \frac{1}{2} \int_0^t S_s \sigma^2 ds.$$

In conclusion

$$S_t = x_0 + \int_0^t S_s \mu ds + \int_0^t S_s \sigma dW_s.$$

Remark 3.4.11 We could have obtained the same result (exercise) by applying Itô formula to $S_t = \phi(Z_t)$, with $Z_t = (\mu - \sigma^2/2)t + \sigma W_t$ (which is an Itô process) and $\phi(x) = x_0 \exp(x)$.

We have just proved the existence of a solution to equation (3.8). We are about to prove its uniqueness. To do that, we shall use the *integration by parts* formula.

Proposition 3.4.12 (integration by parts formula) *Let X_t and Y_t be two Itô processes, $X_t = X_0 + \int_0^t K_s ds + \int_0^t H_s dW_s$ and $Y_t = Y_0 + \int_0^t K'_s ds + \int_0^t H'_s dW_s$. Then*

$$X_t Y_t = X_0 Y_0 + \int_0^t X_s dY_s + \int_0^t Y_s dX_s + \langle X, Y \rangle_t$$

with the following convention

$$\langle X, Y \rangle_t = \int_0^t H_s H'_s ds.$$

Proof. By Itô formula

$$
\begin{aligned}
(X_t + Y_t)^2 &= (X_0 + Y_0)^2 \\
&\quad + 2 \int_0^t (X_s + Y_s) d(X_s + Y_s) \\
&\quad + \int_0^t (H_s + H'_s)^2 ds \\
X_t^2 &= X_0^2 + 2 \int_0^t X_s dX_s + \int_0^t H_s^2 ds \\
Y_t^2 &= Y_0^2 + 2 \int_0^t Y_s dY_s + \int_0^t {H'_s}^2 ds.
\end{aligned}
$$

By subtracting equalities 2 and 3 from the first one, it turns out that

$$X_t Y_t = X_0 Y_0 + \int_0^t X_s dY_s + \int_0^t Y_s dX_s + \int_0^t H_s H'_s ds.$$

$\square$

We now have the tools to show that equation (3.8) has a unique solution. Recall that

$$S_t = x_0 \exp\left(\left(\mu - \sigma^2/2 \right) t + \sigma W_t \right)$$

is a solution of (3.8) and assume that $(X_t)_{t \geq 0}$ is another one. We attempt to compute the *stochastic differential* of the quantity $X_t S_t^{-1}$. Define

$$Z_t = \frac{S_0}{S_t} = \exp\left(\left(-\mu + \sigma^2/2 \right) t - \sigma W_t \right),$$

$\mu' = -\mu + \sigma^2$ and $\sigma' = -\sigma$. Then $Z_t = \exp((\mu' - \sigma'^2/2)t + \sigma' W_t)$ and the verification that we have just done shows that

$$Z_t = 1 + \int_0^t Z_s(\mu' ds + \sigma' dW_s) = 1 + \int_0^t Z_s \left(\left(-\mu + \sigma^2 \right) ds - \sigma dW_s \right).$$

From the integration by parts formula, we can compute the *differential* of $X_t Z_t$

$$d(X_t Z_t) = X_t dZ_t + Z_t dX_t + d\langle X, Z \rangle_t.$$

In this case, we have

$$\langle X, Z \rangle_t = \left\langle \int_0^{\cdot} X_s \sigma dW_s, - \int_0^{\cdot} Z_s \sigma dW_s \right\rangle_t = - \int_0^t \sigma^2 X_s Z_s ds.$$

Therefore

$$\begin{aligned} d(X_t Z_t) &= X_t Z_t \left(\left(-\mu + \sigma^2 \right) dt - \sigma dW_t \right) \\ &+ X_t Z_t \left(\mu dt + \sigma dW_t \right) - X_t Z_t \sigma^2 dt = 0. \end{aligned}$$

Hence, $X_t Z_t$ is equal to $X_0 Z_0$, which implies that

$$\forall t \geq 0, \quad \mathbf{P} \text{ a.s. } X_t = x_0 Z_t^{-1} = S_t.$$

The processes X_t and Z_t being continuous, this proves that

$$\mathbf{P} \text{ a.s. } \forall t \geq 0, \ X_t = x_0 Z_t^{-1} = S_t.$$

We have just proved the following theorem:

Theorem 3.4.13 *If we consider two real numbers σ, μ and a Brownian motion $(W_t)_{t \geq 0}$ and a strictly positive constant T, there exists a unique Itô process $(S_t)_{0 \leq t \leq T}$ which satisfies, for any $t \leq T$,*

$$S_t = x_0 + \int_0^t S_s \left(\mu ds + \sigma dW_s \right).$$

This process is given by

$$S_t = x_0 \exp \left(\left(\mu - \sigma^2/2 \right) t + \sigma W_t \right).$$

Remark 3.4.14

- The process S_t that we just studied will model the evolution of a stock price in the Black-Scholes model.

- When $\mu = 0$, S_t is actually a martingale (see Proposition 3.3.3) called the *exponential martingale of Brownian motion*.

Remark 3.4.15 Let Θ be an open set in $\mathbb{R}$ and $(X_t)_{0 \leq t \leq T}$ an Itô process which stays in Θ at all times. If we consider a function f from Θ to $\mathbb{R}$ which is twice continuously differentiable, we can derive an extension of Itô formula in that case

$$f(X_t) = f(X_0) + \int_0^t f'(X_s) dX_s + \frac{1}{2} \int_0^t f''(X_s) H_s^2 ds.$$

This result allows us to apply Itô formula to $\log(X_t)$ for instance, if X_t is a strictly positive process.

3.4.4 Multidimensional Itô formula

We apply a multidimensional version of Itô formula when f is a function of several Itô processes which are themselves functions of several Brownian motions. This version will prove to be very useful when we model complex interest rate structures for instance.

Definition 3.4.16 *We call standard p-dimensional $\mathcal{F}_t$-Brownian motion an $\mathbb{R}^p$-valued process $(W_t = (W_t^1, \ldots, W_t^p))_{t\geq 0}$ adapted to $\mathcal{F}_t$, where all the $(W_t^i)_{t\geq 0}$ are independent standard $\mathcal{F}_t$-Brownian motions.*

Along these lines, we can define a multidimensional Itô process.

Definition 3.4.17 *$(X_t)_{0\leq t\leq T}$ is an Itô process if*

$$X_t = X_0 + \int_0^t K_s ds + \sum_{i=1}^p \int_0^t H_s^i dW_s^i$$

where:

- *K_t and all the processes (H_t^i) are adapted to $(\mathcal{F}_t)$.*
- *$\int_0^T |K_s| ds < +\infty$ **P** a.s.*
- *$\int_0^T \left(H_s^i\right)^2 ds < +\infty$ **P** a.s.*

Itô formula becomes:

Proposition 3.4.18 *Let $(X_t^1, \ldots, X_t^n)$ be n Itô processes*

$$X_t^i = X_0^i + \int_0^t K_s^i ds + \sum_{j=1}^p \int_0^t H_s^{i,j} dW_s^j$$

then, if f is twice differentiable with respect to x and once differentiable with respect to t, with continuous partial derivatives in (t, x)

$$\begin{aligned}
f(t, X_t^1, \ldots, X_t^n) &= f(0, X_0^1, \ldots, X_0^n) + \int_0^t \frac{\partial f}{\partial s}(s, X_s^1, \ldots, X_s^n) ds \\
&+ \sum_{i=1}^n \int_0^t \frac{\partial f}{\partial x_i}(s, X_s^1, \ldots, X_s^n) dX_s^i \\
&+ \frac{1}{2} \sum_{i,j=1}^n \int_0^t \frac{\partial^2 f}{\partial x_i x_j}(s, X_s^1, \ldots, X_s^n) d\langle X^i, X^j \rangle_s
\end{aligned}$$

with:

- *$dX_s^i = K_s^i ds + \sum_{j=1}^p H_s^{i,j} dW_s^j$,*
- *$d\langle X^i, X^j \rangle_s = \sum_{m=1}^p H_s^{i,m} H_s^{j,m} ds$.*

Remark 3.4.19 If $(X_s)_{0\leq t\leq T}$ and $(Y_s)_{0\leq t\leq T}$ are two Itô processes, we can define formally the *cross-variation* of X and Y (denoted by $\langle X, Y \rangle_s$) through the following properties:

- $\langle X, Y \rangle_t$ is bilinear and symmetric.
- $\langle \int_0^{\cdot} K_s ds, X_{\cdot} \rangle_t = 0$ if $(X_t)_{0\leq t\leq T}$ is an Itô process.

- $\langle \int_0^{\cdot} H_s dW_t^i, \int_0^{\cdot} H_s' dW_t^j \rangle_t = 0$ if $i \neq j$.
- $\langle \int_0^{\cdot} H_s dW_t^i, \int_0^{\cdot} H_s' dW_t^i \rangle_t = \int_0^t H_s H_s' ds$.

This definition leads to the cross-variation stated in the previous proposition.

3.5 Stochastic differential equations

In Section 3.4.2, we studied in detail the solutions to the equation

$$X_t = x + \int_0^t X_s (\mu ds + \sigma dW_s).$$

We can now consider some more general equations of the type

$$X_t = Z + \int_0^t b(s, X_s) ds + \int_0^t \sigma(s, X_s) dW_s. \tag{3.10}$$

These equations are called *stochastic differential equations* and a solution of (3.10) is called a *diffusion*. These equations are useful to model most financial assets, whether we are speaking about stocks or interest rate processes. Let us first study some properties of the solutions to these equations.

3.5.1 Itô theorem

What do we mean by *a solution of (3.10)*?

Definition 3.5.1 *We consider a probability space* $(\Omega, \mathcal{A}, \mathbf{P})$ *equipped with a filtration* $(\mathcal{F}_t)_{t \geq 0}$. *We also have functions* $b : \mathbb{R}^+ \times \mathbb{R} \to \mathbb{R}$, $\sigma : \mathbb{R}^+ \times \mathbb{R} \to \mathbb{R}$, *a* $\mathcal{F}_0$-*measurable random variable* Z *and finally an* $\mathcal{F}_t$-*Brownian motion* $(W_t)_{t \geq 0}$. *A solution to equation (3.10) is an* $\mathcal{F}_t$-*adapted stochastic process* $(X_t)_{t \geq 0}$ *that satisfies:*

- *For any* $t \geq 0$, *the integrals* $\int_0^t b(s, X_s) ds$ *and* $\int_0^t \sigma(s, X_s) dW_s$ *exist*

$$\int_0^t |b(s, X_s)| ds < +\infty \ and \ \int_0^t |\sigma(s, X_s)|^2 ds < +\infty \ \mathbf{P} \ a.s.$$

- $(X_t)_{t \geq 0}$ *satisfies (3.10), i.e.*

$$\forall t \geq 0 \ \mathbf{P} \ a.s. \ X_t = Z + \int_0^t b(s, X_s) \, ds + \int_0^t \sigma(s, X_s) \, dW_s.$$

Remark 3.5.2 Formally, we often write equation (3.10) as

$$\begin{cases} dX_t &= b(t, X_t) \, dt + \sigma(t, X_t) \, dW_t \\ X_0 &= Z \end{cases}$$

The following theorem gives sufficient conditions on b and σ to guarantee the existence and uniqueness of a solution of equation (3.10).

Theorem 3.5.3 *If* b *and* σ *are continuous functions and if there exists a constant* $K < +\infty$ *such that*

1. $|b(t,x) - b(t,y)| + |\sigma(t,x) - \sigma(t,y)| \leq K|x - y|$
2. $|b(t,x)| + |\sigma(t,x)| \leq K(1 + |x|)$
3. $\mathbf{E}(Z^2) < +\infty$

then, for any $T \geq 0$, (3.10) admits a unique solution in the interval $[0,T]$. Moreover, this solution $(X_s)_{0 \leq s \leq T}$ satisfies

$$\mathbf{E}\left(\sup_{0 \leq s \leq T} |X_s|^2\right) < +\infty$$

The uniqueness of the solution means that if $(X_t)_{0 \leq t \leq T}$ and $(Y_t)_{0 \leq t \leq T}$ are two solutions of (3.10), then $\mathbf{P}$ a.s. $\forall 0 \leq t \leq T$, $X_t = Y_t$.

Proof.   We define the set

$$\mathcal{E} = \Big\{ (X_s)_{0 \leq s \leq T}, \ \mathcal{F}_t\text{-adapted continuous processes,}$$

$$\text{such that } \mathbf{E}\left(\sup_{s \leq T} |X_s|^2\right) < +\infty \Big\}.$$

Together with the norm $\|X\| = \left(\mathbf{E}\left(\sup_{0 \leq s \leq T} |X_s|^2\right)\right)^{1/2} \mathcal{E}$ is a complete normed vector space. In order to show the existence of a solution, we are going to use the theorem of existence of a fixed point for a contracting mapping. Let Φ be the function that maps a process $(X_s)_{0 \leq s \leq T}$ into a process $(\Phi(X)_s)_{0 \leq s \leq T}$ defined by

$$\Phi(X)_t = Z + \int_0^t b(s, X_s)ds + \int_0^t \sigma(s, X_s)dW_s.$$

If X belongs to $\mathcal{E}$, $\Phi(X)$ is well defined and furthermore if X and Y are both in $\mathcal{E}$ we can use the fact that $(a + b)^2 \leq 2(a^2 + b^2)$ to write that

$$|\Phi(X)_t - \Phi(Y)_t|^2 \leq 2\left(\sup_{0 \leq t \leq T}\left|\int_0^t (b(s, X_s) - b(s, Y_s))ds\right|^2\right.$$
$$\left. + \sup_{0 \leq t \leq T}\left|\int_0^t (\sigma(s, X_s) - \sigma(s, Y_s))dW_s\right|^2\right)$$

and therefore by inequality (3.4)

$$\mathbf{E}\left(\sup_{s \leq T} |\Phi(X)_t - \Phi(Y)_t|^2\right)$$

$$\leq 2\mathbf{E}\left(\sup_{0 \leq t \leq T}\left(\int_0^t |b(s, X_s) - b(s, Y_s)|ds\right)^2\right)$$

$$+ 8\mathbf{E}\left(\int_0^T (\sigma(s, X_s) - \sigma(s, Y_s))^2 ds\right)$$

$$\leq 2(K^2 T^2 + 4K^2 T)\mathbf{E}\left(\sup_{0 \leq t \leq T} |X_t - Y_t|^2\right)$$

whence $\|\Phi(X) - \Phi(Y)\| \leq (2(K^2T^2 + 4K^2T))^{1/2} \, \|X - Y\|$. On the other hand, if we denote by 0 the process that is identically equal to zero and if we notice that $(a + b + c)^2 \leq 3(a^2 + b^2 + c^2)$

$$|\Phi(0)_t|^2 \leq 3 \left(Z^2 + \sup_{0 \leq t \leq T} \left| \int_0^t b(s, 0)ds \right|^2 + \sup_{0 \leq t \leq T} \left| \int_0^t \sigma(s, 0)dW_s \right|^2 \right)$$

Therefore

$$\mathbf{E} \left(\sup_{0 \leq t \leq T} |\Phi(0)_t|^2 \right) \leq 3(\mathbf{E}(Z^2) + K^2T^2 + 4K^2T) < +\infty.$$

We deduce that Φ is a mapping from $\mathcal{E}$ to $\mathcal{E}$ with a Lipschitz norm bounded from above by $k(T) = (2(K^2T^2 + 4K^2T))^{1/2}$. If we assume that T is small enough so that $k(T) < 1$, it turns out that Φ is a contraction from $\mathcal{E}$ into $\mathcal{E}$. Thus it has a fixed point in $\mathcal{E}$. Moreover, if X is a fixed point of Φ, it is a solution of (3.10). That completes the proof of the existence for T small enough. On the other hand, a solution of (3.10) which belongs to $\mathcal{E}$ is a fixed point of Φ. That proves the uniqueness of a solution of equation (3.10) in the class $\mathcal{E}$. In order to prove the uniqueness in the whole class of Itô processes, we just need to show that a solution of (3.10) always belongs to $\mathcal{E}$. Let X be a solution of (3.10), and define $T_n = \inf\{s \geq 0, |X_s| > n\}$ and $f^n(t) = \mathbf{E} \left(\sup_{0 \leq s \leq t \wedge T_n} |X_s|^2 \right)$. It is easy to check that $f^n(t)$ is finite and continuous. Using the same comparison arguments as before, we can say that

$$\begin{aligned}
\mathbf{E} \left(\sup_{0 \leq u \leq t \wedge T_n} |X_u|^2 \right) &\leq 3 \left(\mathbf{E}(Z^2) + \mathbf{E} \left(\int_0^{t \wedge T_n} K(1 + |X_s|)ds \right)^2 \right. \\
&\quad \left. + 4\mathbf{E} \left(\int_0^{t \wedge T_n} K^2(1 + |X_s|)^2 ds \right) \right) \\
&\leq 3 \left(\mathbf{E}(Z^2) + 2(K^2T + 4K^2) \right. \\
&\quad \left. \times \int_0^t \left(1 + \mathbf{E} \left(\sup_{0 \leq u \leq s \wedge T_n} |X_u|^2 \right) \right) ds \right).
\end{aligned}$$

This yields the following inequality

$$f^n(t) \leq a + b \int_0^t f^n(s)ds.$$

In order to complete the proof, let us recall the Gronwall lemma.

Lemma 3.5.4 (Gronwall lemma) *If f is a continuous function such that for any $0 \leq t \leq T$, $f(t) \leq a + b \int_0^t f(s)ds$, then $f(T) \leq a(1 + e^{bT})$.*

Proof. Let us write $u(t) = e^{-bt} \int_0^t f(s)ds$. Then,

$$u'(t) = e^{-bt}(f(s) - b \int_0^t f(s)ds) \leq ae^{-bt}.$$

By first-order integration we obtain $u(T) \leq a/b$ and $f(T) \leq a(1 + e^{bT})$. $\square$

In our case, we have $f^n(T) < K < +\infty$, where K is a function of T independent of n. It follows from Fatou lemma that, for any T,

$$\mathbf{E}\left(\sup_{0\leq s\leq T}|X_s|^2\right) < K < +\infty.$$

Therefore X belongs to $\mathcal{E}$ and that completes the proof for small T. For an arbitrary T, we consider a large enough integer n and think successively on the intervals $[0, T/n], [T/n, 2T/n], \ldots, [(n-1)T/n, T]$. $\qquad\square$

3.5.2 The Ornstein-Ulhenbeck process

Ornstein-Ulhenbeck process is the unique solution of the following equation:

$$\begin{cases} dX_t &= -cX_t dt + \sigma dW_t \\ X_0 &= x \end{cases}$$

It can be written explicitly. Indeed, if we consider $Y_t = X_t e^{ct}$ and integrate by parts, it yields

$$dY_t = dX_t e^{ct} + X_t d(e^{ct}) + d\langle X, e^{c\cdot}\rangle_t.$$

Furthermore $\langle X, e^{c\cdot}\rangle_t = 0$ because $d(e^{ct}) = ce^{ct}dt$. It follows that $dY_t = \sigma e^{ct}dW_t$ and thus

$$X_t = xe^{-ct} + \sigma e^{-ct}\int_0^t e^{cs}dW_s.$$

This enables us to compute the mean and variance of X_t:

$$\mathbf{E}(X_t) = xe^{-ct} + \sigma e^{-ct}\mathbf{E}\left(\int_0^t e^{cs}dW_s\right) = xe^{-ct}$$

(since $\mathbf{E}\left(\int_0^t (e^{cs})^2 ds\right) < +\infty$, $\int_0^t e^{cs}dW_s$ is a martingale null at time 0 and therefore its expectation is zero). Similarly

$$\begin{aligned} \mathrm{Var}(X_t) &= \mathbf{E}\left((X_t - \mathbf{E}(X_t))^2\right) \\ &= \sigma^2\mathbf{E}\left(e^{-2ct}\left(\int_0^t e^{cs}dW_s\right)^2\right) \\ &= \sigma^2 e^{-2ct}\mathbf{E}\left(\int_0^t e^{2cs}ds\right) \\ &= \sigma^2\frac{1-e^{-2ct}}{2c}. \end{aligned}$$

We can also prove that X_t is a normal random variable, since X_t can be written as $\int_0^t f(s)dW_s$ where $f(.)$ is a deterministic function of time and $\int_0^t f^2(s)ds < +\infty$ (see Exercise 12). More precisely, the *process* $(X_t)_{t\geq 0}$ is Gaussian. This means that if $\lambda_1, \ldots, \lambda_n$ are real numbers and if $0 \le t_1 < \cdots < t_n$, the random variable $\lambda_1 X_{t_1} + \cdots + \lambda_n X_{t_n}$ is normal. To convince ourselves, we just notice that

$$X_{t_i} = x e^{-ct_i} + \sigma e^{-ct_i} \int_0^{+\infty} 1_{\{s \le t_i\}} e^{cs} dW_s = m_i + \int_0^t f_i(s) dW_s.$$

Then $\lambda_1 X_{t_1} + \cdots + \lambda_n X_{t_n} = \sum_{i=1}^n \lambda_i m_i + \int_0^t \left(\sum_{i=1}^n \lambda_i f_i(s) \right) dW_s$ which is indeed a normal random variable (since it is a stochastic integral of a deterministic function of time).

3.5.3 Multidimensional stochastic differential equations

The analysis of stochastic differential equations can be extended to the case when processes evolve in $\mathbb{R}^n$. This generalisation proves to be useful in finance when we want to model baskets of stocks or currencies. We consider

- $W = (W^1, \ldots, W^p)$ an $\mathbb{R}^p$-valued $\mathcal{F}_t$-Brownian motion.
- $b : \mathbb{R}^+ \times \mathbb{R}^n \to \mathbb{R}^n$, $b(s, x) = (b^1(s, x), \ldots, b^n(s, x))$.
- $\sigma : \mathbb{R}^+ \times \mathbb{R}^n \to \mathbb{R}^{n \times p}$ (which is the set of $n \times p$ matrices),

$$\sigma(s, x) = (\sigma_{i,j}(s, x))_{1 \le i \le n, 1 \le j \le p}.$$

- $Z = (Z^1, \ldots, Z^n)$ an $\mathcal{F}_0$-measurable random variable in $\mathbb{R}^n$.

We are also interested in the following stochastic differential equation:

$$X_t = Z + \int_0^t b(s, X_s) \, ds + \int_0^t \sigma(s, X_s) \, dW_s. \qquad (3.11)$$

In other words, we are looking for a process $(X_t)_{0 \le t \le T}$ with values in $\mathbb{R}^n$, adapted to the filtration $(\mathcal{F}_t)_{t \ge 0}$ and such that $\mathbf{P}$ a.s. , for any t and for any $i \le n$

$$X_t^i = Z^i + \int_0^t b^i(s, X_s) ds + \sum_{j=1}^p \int_0^t \sigma_{i,j}(s, X_s) dW_s^j.$$

The theorem of existence and uniqueness of a solution of (3.11) can be stated as:

Theorem 3.5.5 *If $x \in \mathbb{R}^n$, we denote by $|x|$ the Euclidean norm of x and if $\sigma \in \mathbb{R}^{n \times p}$, $|\sigma|^2 = \sum_{1 \le i \le n, 1 \le j \le p} \sigma_{i,j}^2$. We assume that*

1. $|b(t, x) - b(t, y)| + |\sigma(t, x) - \sigma(t, y)| \le K|x - y|$
2. $|b(t, x)| + |\sigma(t, x)| \le K(1 + |x|)$
3. $\mathbf{E}(|Z|^2) < +\infty$

then, there exists a unique solution to (3.11). Moreover, this solution satisfies for any T

$$\mathbf{E} \left(\sup_{0 \le s \le T} |X_s|^2 \right) < +\infty.$$

The proof is very similar to the one in the scalar case.

3.5.4 The Markov property of the solution of a stochastic differential equation

The intuitive meaning of the Markov property is that the future behaviour of the process $(X_t)_{t\geq 0}$ after t depends only on the value X_t and is not influenced by the history of the process before t. This is a crucial property of the Markovian model and it will have great consequences in the pricing of options. For instance, it will allow us to show that the price of an option on an underlying asset whose price is Markovian depends only on the price of this underlying asset at time t.

Mathematically speaking, an $\mathcal{F}_t$-adapted process $(X_t)_{t\geq 0}$ satisfies the Markov property if, for any bounded Borel function f and for any s and t such that $s \leq t$, we have

$$\mathbf{E}\left(f\left(X_t\right)|\mathcal{F}_s\right) = \mathbf{E}\left(f\left(X_t\right)|X_s\right).$$

We are going to state this property for a solution of equation (3.10). We shall denote by $(X_s^{t,x}, s \geq t)$ the solution of equation (3.10) starting from x at time t and by $X^x = X^{0,x}$ the solution starting from x at time 0. For $s \geq t$, $X_s^{t,x}$ satisfies

$$X_s^{t,x} = x + \int_t^s b\left(u, X_u^{t,x}\right) du + \int_t^s \sigma\left(u, X_u^{t,x}\right) dW_u.$$

A priori, $X^{t,x}$ is defined for any (t, x) almost surely. However, under the assumptions of Theorem 3.5.3, we can build a process depending on (t, x, s) which is almost surely continuous with respect to these variables and is a solution of the previous equation. This result is difficult to prove and the interested reader should refer to Rogers and Williams (1987) for the proof.

The Markov property is a consequence of the *flow* property of a solution of a stochastic differential equation which is itself an extension of the flow property of solutions of ordinary differential equations.

Lemma 3.5.6 *Under the assumptions of Theorem 3.5.3, if $s \geq t$*

$$X_s^{0,x} = X_s^{t,X_t^x} \quad \mathbf{P} \ a.s.$$

Proof. We are only going to sketch the proof of this lemma. For any x, we have

$$\mathbf{P} \ a.s. \ X_s^{t,x} = x + \int_t^s b\left(u, X_u^{t,x}\right) du + \int_t^s \sigma\left(u, X_u^{t,x}\right) dW_u.$$

It follows that, $\mathbf{P}$ a.s. for any $y \in \mathbb{R}$,

$$X_s^{t,y} = y + \int_t^s b\left(u, X_u^{t,y}\right) du + \int_t^s \sigma\left(u, X_u^{t,y}\right) dW_u,$$

and also

$$X_s^{t,X_t^x} = X_t^x + \int_t^s b\left(u, X_u^{t,X_t^x}\right) du + \int_t^s \sigma\left(u, X_u^{t,X_t^x}\right) dW_u.$$

These results are intuitive, but they can be proved rigorously by using the continuity of $y \mapsto X^{t,y}$. We can also notice that X_s^x is also a solution of the previous equation.

Indeed, if $t \leq s$

$$
\begin{aligned}
X_s^{0,x} &= x + \int_0^s b\left(u, X_u^x\right) du + \int_0^s \sigma\left(u, X_u^x\right) dW_u \\
&= X_t^x + \int_t^s b\left(u, X_u^x\right) du + \int_t^s \sigma\left(u, X_u^x\right) dW_u.
\end{aligned}
$$

The uniqueness of the solution to this equation implies that $X_s^{0,x} = X_s^{t,X_t}$ for $t \leq s$. □

In this case, the Markov property can be stated as follows:

Theorem 3.5.7 *Let* $(X_t)_{t \geq 0}$ *be a solution of (3.10). It is a Markov process with respect to the Brownian filtration* $(\mathcal{F}_t)_{t \geq 0}$*. Furthermore, for any bounded Borel function* f *we have*

$$
\mathbf{P} \text{ a.s. } \mathbf{E}\left(f\left(X_t\right) | \mathcal{F}_s\right) = \phi(X_s),
$$

with $\phi(x) = \mathbf{E}\left(f(X_t^{s,x})\right)$.

Remark 3.5.8 The previous equality is often written as

$$
\mathbf{E}\left(f\left(X_t\right) | \mathcal{F}_s\right) = \mathbf{E}\left(f(X_t^{s,x})\right)|_{x=X_s}.
$$

Proof. Yet again, we shall only sketch the proof of this theorem. For a full proof, the reader ought to refer to Friedman (1975).

The flow property shows that, if $s \leq t$, $X_t^x = X_t^{s,X_s^x}$. On the other hand, we can prove that $X_t^{s,x}$ is a measurable function of x and the Brownian increments $(W_{s+u} - W_s, \ u \geq 0)$ (this result is natural but it is quite tricky to justify (see Friedman (1975)). If we use this result for fixed s and t we obtain $X_t^{s,x} = \Phi(x, W_{s+u} - W_s; \ u \geq 0)$ and thus

$$
X_t^x = \Phi(X_s^x, W_{s+u} - W_s; \ u \geq 0),
$$

where X_s^x is $\mathcal{F}_s$-measurable and $(W_{s+u} - W_s)_{u \geq 0}$ is independent of $\mathcal{F}_s$.

If we apply the result of Proposition A.2.5 in the Appendix to X_s, $(W_{s+u} - W_s)_{u \geq 0}$, Φ and $\mathcal{F}_s$, it turns out that

$$
\begin{aligned}
\mathbf{E}\left(f\left(\Phi(X_s^x, W_{s+u} - W_s; \ u \geq 0)\right) | \mathcal{F}_s\right) \\
= \mathbf{E}\left(f\left(\Phi(x, W_{s+u} - W_s; \ u \geq 0)\right)\right)|_{x=X_s^x} \\
= \mathbf{E}\left(f(X_t^{s,x})\right)|_{x=X_s^x}.
\end{aligned}
$$

□

The previous result can be extended to the case when we consider a function of the whole path of a diffusion after time s. In particular, the following theorem is useful when we do computations involving interest rate models.

Theorem 3.5.9 *Let* $(X_t)_{t \geq 0}$ *be a solution of (3.10) and* $r(s, x)$ *be a non-negative* *measurable function. For $t > s$*

$$\mathbf{P} \text{ a.s. } \mathbf{E}\left(e^{-\int_s^t r(u, X_u)du} f(X_t) \,|\, \mathcal{F}_s\right) = \phi(X_s)$$

with

$$\phi(x) = \mathbf{E}\left(e^{-\int_s^t r(u, X_u^{s,x})du} f(X_t^{s,x})\right).$$

It is also written as

$$\mathbf{E}\left(e^{-\int_s^t r(u, X_u)du} f(X_t) \,|\, \mathcal{F}_s\right) = \mathbf{E}\left(e^{-\int_s^t r(u, X_u^{s,x})du} f(X_t^{s,x})\right)\Bigg|_{x=X_s}.$$

Remark 3.5.10 Actually, one can prove a more general result. Without getting into the technicalities, let us just mention that if ϕ is *a function of the whole path* of X_t after time s, the following stronger result is still true:

$$\mathbf{P} \text{ a.s. } \mathbf{E}\left(\phi\left(X_t^x, \, t \geq s\right) \,|\, \mathcal{F}_s\right) = \mathbf{E}\left(\phi(X_t^{s,x}, \, t \geq s)\right)\big|_{x=X_s}.$$

Remark 3.5.11 When b and σ are independent of x (the diffusion is said to be homogeneous), we can show that the law of $X_{s+t}^{s,x}$ is the same as the one of $X_t^{0,x}$, which implies that if f is a bounded measurable function, then

$$\mathbf{E}\left(f(X_{s+t}^{s,x})\right) = \mathbf{E}\left(f(X_t^{0,x})\right).$$

We can extend this result and show that if r is a function of x only then

$$\mathbf{E}\left(e^{-\int_s^{s+t} r(X_u^{s,x})du} f(X_{s+t}^{s,x})\right) = \mathbf{E}\left(e^{-\int_0^t r(X_u^{0,x})du} f(X_t^{0,x})\right).$$

In that case, the Theorem 3.5.9 becomes

$$\mathbf{E}\left(e^{-\int_s^t r(X_u)du} f(X_t) \,|\, \mathcal{F}_s\right) = \mathbf{E}\left(e^{-\int_0^{t-s} r(X_u^{0,x})du} f(X_{t-s}^{0,x})\right)\Bigg|_{x=X_s}.$$

3.6 Exercises

Exercise 6 Let $(M_t)_{t\geq0}$ be a martingale such that for any t, $\mathbf{E}(M_t^2) < +\infty$. Prove that if $s \leq t$

$$\mathbf{E}\left((M_t - M_s)^2 \,|\, \mathcal{F}_s\right) = \mathbf{E}\left(M_t^2 - M_s^2 \,|\, \mathcal{F}_s\right).$$

Exercise 7 Let X_t be a process with independent stationary increments and zero initial value such that for any t, $\mathbf{E}\left(X_t^2\right) < +\infty$. We shall also assume that the map $t \mapsto \mathbf{E}\left(X_t^2\right)$ is continuous. Prove that $\mathbf{E}(X_t) = ct$ and that $\mathrm{Var}(X_t) = c't$, where c and c' are two constants.

Exercise 8 Prove that, if τ is a stopping time,

$$\mathcal{F}_\tau = \{A \in \mathcal{A}, \text{ for all } t \geq 0, A \cap \{\tau \leq t\} \in \mathcal{F}_t\}$$

is a σ-algebra.

Exercise 9 Let S be a stopping time, prove that S is $\mathcal{F}_S$-measurable.

Exercise 10 Let S and T be two stopping times such that $S \leq T$ **P** a.s. Prove that $\mathcal{F}_S \subset \mathcal{F}_T$.

Exercise 11 Let S be a stopping time almost surely finite, and $(X_t)_{t\geq 0}$ be an adapted process almost surely continuous.

1. Prove that, **P** a.s., for any s

$$X_s = \lim_{n\to+\infty} \sum_{k\geq 0} \mathbf{1}_{[k/n,(k+1)/n[}(s) X_{k/n}(\omega).$$

2. Prove that the mapping

$$([0,t] \times \Omega, \mathcal{B}([0,t]) \times \mathcal{F}_t) \longrightarrow (\mathbb{R}, \mathcal{B}(\mathbb{R}))$$
$$(s, \omega) \longmapsto X_s(\omega)$$

   is measurable.

3. Conclude that if $S \leq t$, X_S is $\mathcal{F}_t$-measurable, and thus that X_S is $\mathcal{F}_S$-measurable.

Exercise 12 This exercise is an introduction to the concept of stochastic integration. We want to build an integral of the form $\int_0^{+\infty} f(s)dX_s$, where $(X_t)_{t\geq 0}$ is an $\mathcal{F}_t$-Brownian motion and $f(s)$ is a measurable function from $(\mathbb{R}^+, \mathcal{B}(\mathbb{R}^+))$ into $(\mathbb{R}, \mathcal{B}(\mathbb{R}))$ such that $\int_0^{+\infty} f^2(s)ds < +\infty$. This type of integral is called *Wiener integral* and it is a particular case of Itô integral which is studied in Section 3.4.

We recall that the set $\mathcal{H}$ of functions of the form $\sum_{0\leq i\leq N-1} a_i \mathbf{1}_{]t_i,t_{i+1}]}$, with $a_i \in \mathbb{R}$, and $t_0 = 0 \leq t_1 \leq \cdots \leq t_N$ is dense in the space $L^2(\mathbb{R}^+, dx)$ endowed with the norm $\|f\|_{L^2} = \left(\int_0^{+\infty} f^2(s)ds\right)^{1/2}$.

1. Consider $a_i \in \mathbb{R}$, and $0 = t_0 \leq t_1 \leq \cdots \leq t_N$, and call

$$f = \sum_{0\leq i\leq N-1} a_i \mathbf{1}_{]t_i,t_{i+1}]}.$$

   We define

$$I_e(f) = \sum_{0\leq i\leq N-1} a_i(X_{t_{i+1}} - X_{t_i}).$$

   Prove that $I_e(f)$ is a normal random variable and compute its mean and variance. In particular, show that

$$\mathbf{E}(I_e(f)^2) = \|f\|_{L^2}^2.$$

2. From this, show that there exists a unique linear mapping I from $L^2(\mathbb{R}^+, dx)$ into $L^2(\Omega, \mathcal{F}, \mathbf{P})$, such that $I(f) = I_e(f)$, when f belongs to $\mathcal{H}$ and $\mathbf{E}(I(f)^2) = \|f\|_{L^2}$, for any f in $L^2(\mathbb{R}^+)$.

3. Prove that if $(X_n)_{n\geq 0}$ is a sequence of normal random variables with zero-mean which converge to X in $L^2(\Omega, \mathcal{F}, \mathbf{P})$, then X is also a normal random

variable with zero-mean. Deduce that if $f \in L^2(\mathbb{R}^+, dx)$ then $I(f)$ is a normal random variable with zero mean and a variance equal to $\int_0^{+\infty} f^2(s)ds$.

4. We consider $f \in L^2(\mathbb{R}^+, dx)$, and we define

$$Z_t = \int_0^t f(s)dX_s = \int 1_{]0,t]}(s)f(s)dX_s,$$

prove that Z_t is adapted to $\mathcal{F}_t$, and that $Z_t - Z_s$ is independent of $\mathcal{F}_s$ (hint: begin with the case $f \in H$).

5. Prove that the processes Z_t, $Z_t^2 - \int_0^t f^2(s)ds$, $\exp(Z_t - \frac{1}{2}\int_0^t f^2(s)ds)$ are $\mathcal{F}_t$-martingales.

Exercise 13 Let T be a positive real number and $(M_t)_{0 \leq t \leq T}$ be a continuous $\mathcal{F}_t$-martingale. We assume that $\mathbf{E}(M_T^2)$ is finite.

1. Prove that $(|M_t|)_{0 \leq t \leq T}$ is a submartingale.

2. Show that, if $M^* = \sup_{0 \leq t \leq T} |M_t|$,

$$\lambda \mathbf{P}(M^* \geq \lambda) \leq \mathbf{E}\left(|M_T|1_{\{M^* \geq \lambda\}}\right).$$

(Hint: apply the optional sampling theorem to the submartingale $|M_t|$ between $\tau \wedge T$ and T where $\tau = \inf\{t \leq T, |M_t| \geq \lambda\}$ (if this set is empty τ is equal to $+\infty$).)

3. From the previous result, deduce that for positive A

$$\mathbf{E}((M^* \wedge A)^2) \leq 2\mathbf{E}((M^* \wedge A)|M_T|).$$

(Use the fact that $(M^* \wedge A)^p = \int_0^{M^* \wedge A} px^{p-1}dx$ for $p = 1, 2$.)

4. Prove that $\mathbf{E}(M^*)$ is finite and

$$\mathbf{E}\left(\sup_{0 \leq t \leq T} |M_t|^2\right) \leq 4\mathbf{E}(|M_T|^2).$$

Exercise 14

1. Prove that if S and S' are two $\mathcal{F}_t$-stopping times then $S \wedge S' = \inf(S, S')$ and $S \vee S' = \sup(S, S')$ are also two $\mathcal{F}_t$-stopping times.

2. By applying the sampling theorem to the stopping time $S \vee s$ prove that

$$\mathbf{E}\left(M_S 1_{\{S>s\}}|\mathcal{F}_s\right) = M_s 1_{\{S>s\}}.$$

3. Deduce that for $s \leq t$

$$\mathbf{E}\left(M_{S \wedge t} 1_{\{S>s\}}|\mathcal{F}_s\right) = M_s 1_{\{S>s\}}.$$

4. Remembering that $M_{S \wedge s}$ is $\mathcal{F}_s$-measurable, show that $t \to M_{S \wedge t}$ is an $\mathcal{F}_t$-martingale.

Exercise 15

1. Let $(H_t)_{0 \leq t \leq T}$ be an adapted measurable process such that $\int_0^T H_t^2 dt < \infty$, a.s. Let $M_t = \int_0^t H_s dW_s$. Show that if $\mathbf{E}\left(\sup_{0 \leq t \leq T} M_t^2\right) < \infty$, then $\mathbf{E}\left(\int_0^T H_t^2 dt\right) < \infty$. Hint: introduce the sequence of stopping times $\tau_n = \inf\{t \geq 0 \mid \int_0^t H_s^2 ds = n\}$ and show that $E(M_{T \wedge \tau_n}^2) = \mathbf{E}\left(\int_0^{T \wedge \tau_n} H_s^2 ds\right)$.

2. Let $p(t, x) = 1/\sqrt{1 - t} \exp(-x^2/2(1 - t))$, for $0 \leq t < 1$ and $x \in \mathbb{R}$, and $p(1, x) = 0$. Define $M_t = p(t, W_t)$, where $(W_t)_{0 \leq t \leq 1}$ is standard Brownian motion.

   (a) Prove that
   $$M_t = M_0 + \int_0^t \frac{\partial p}{\partial x}(s, W_s) dW_s.$$

   (b) Let
   $$H_t = \frac{\partial p}{\partial x}(t, W_t).$$
   Prove that $\int_0^1 H_t^2 dt < \infty$, a.s. and $\mathbf{E}\left(\int_0^1 H_t^2 dt\right) = +\infty$.

Exercise 16 Let $(M_t)_{0 \leq t \leq T}$ be a continuous $\mathcal{F}_t$-martingale equal to $\int_0^t K_s ds$, where $(K_t)_{0 \leq t \leq T}$ is an $\mathcal{F}_t$-adapted process such that $\int_0^T |K_s| ds < +\infty$ **P** a.s.

1. Moreover, we assume that **P** a.s. $\int_0^T |K_s| ds \leq C < +\infty$. Prove that if we write $t_i^n = Ti/n$ for $0 \leq i \leq n$, then
   $$\lim_{n \to +\infty} \mathbf{E}\left(\sum_{i=1}^n \left(M_{t_i^n} - M_{t_{i-1}^n}\right)^2\right) = 0.$$

2. Under the same assumptions, prove that
   $$\mathbf{E}\left(\sum_{i=1}^n \left(M_{t_i^n} - M_{t_{i-1}^n}\right)^2\right) = \mathbf{E}\left(M_T^2 - M_0^2\right).$$
   Conclude that $M_T = 0$ **P** a.s. , and thus **P** a.s. $\forall t \leq T$, $M_t = 0$.

3. $\int_0^T |K_s| ds$ is now assumed to be finite almost surely as opposed to bounded. We shall accept the fact that the random variable $\int_0^t |K_s| ds$ is $\mathcal{F}_t$-measurable. Show that T_n defined by
   $$T_n = \inf\{0 \leq s \leq T, \int_0^t |K_s| ds \geq n\}$$
   (or T if this set is empty) is a stopping time. Prove that **P** a.s. $\lim_{n \to +\infty} T_n = T$. Considering the sequence of martingales $(M_{t \wedge T_n})_{t \geq 0}$, prove that
   $$\textbf{P a.s. } \forall t \leq T, \ M_t = 0.$$

4. Let M_t be a martingale of the form $\int_0^t H_s dW_s + \int_0^t K_s ds$ with $\int_0^t H_s^2 ds < +\infty$ **P** a.s. and $\int_0^t |K_s| ds < +\infty$ **P** a.s. Define the sequence of stopping times $T_n = \inf\{t \le T, \int_0^t H_s^2 ds \ge n\}$, in order to prove that $K_t = 0$ $dt \times$ **P** a.s.

Exercise 17 Let us call X_t the solution of the following stochastic differential equation

$$\begin{cases} dX_t & = & (\mu X_t + \mu')dt + (\sigma X_t + \sigma')dW_t \\ X_0 & = & 0. \end{cases}$$

We write $S_t = \exp\left((\mu - \sigma^2/2)t + \sigma W_t\right)$.

1. Derive the stochastic differential equation satisfied by S_t^{-1}.

2. Prove that

$$d(X_t S_t^{-1}) = S_t^{-1}\left((\mu' - \sigma\sigma')dt + \sigma'dW_t\right).$$

3. Obtain the explicit representation of X_t.

Exercise 18 Let $(W_t)_{t \ge 0}$ be an F_t-Brownian motion. The purpose of this exercise is to compute the law of $(W_t, \sup_{s \le t} W_s)$.

1. Consider S a bounded stopping time. Apply the optional sampling theorem to the martingale $M_t = \exp(izW_t + z^2 t/2)$, where z is a real number to prove that if $0 \le u \le v$ then

$$\mathbf{E}\left(\exp\left(iz(W_{v+S} - W_{u+S})\right)|\mathcal{F}_{u+S}\right) = \exp\left(-z^2(v-u)/2\right).$$

2. Deduce that $W_u^S = W_{u+S} - W_S$ is an $\mathcal{F}_{S+u}$-Brownian motion independent of the σ-algebra $\mathcal{F}_S$.

3. Let $(Y_t)_{t \ge 0}$ be a continuous stochastic process independent of the σ-algebra $\mathcal{B}$ such that $\mathbf{E}(\sup_{0 \le s \le K} |Y_s|) < +\infty$. Let T be a non-negative $\mathcal{B}$-measurable random variable bounded from above by K. Show that

$$\mathbf{E}(Y_T|\mathcal{B}) = \mathbf{E}(Y_t)|_{t=T}.$$

We shall start by assuming that T can be written as $\sum_{1 \le i \le n} t_i \mathbf{1}_{A_i}$, where $0 < t_1 < \cdots < t_n = K$, and the A_i are disjoint $\mathcal{B}$-measurable sets.

4. We denote by τ^λ the $\inf\{s \ge 0, W_s > \lambda\}$, prove that if f is a bounded Borel function we have

$$\mathbf{E}\left(f(W_t)\mathbf{1}_{\{\tau^\lambda \le t\}}\right) = \mathbf{E}\left(\mathbf{1}_{\{\tau^\lambda \le t\}}\phi(t - \tau^\lambda)\right),$$

where $\phi(u) = \mathbf{E}(f(W_u + \lambda))$. Notice that $\mathbf{E}(f(W_u + \lambda)) = \mathbf{E}(f(-W_u + \lambda))$ and prove that

$$\mathbf{E}\left(f(W_t)\mathbf{1}_{\{\tau^\lambda \le t\}}\right) = \mathbf{E}\left(f(2\lambda - W_t)\mathbf{1}_{\{\tau^\lambda \le t\}}\right).$$

5. Show that if we write $W_t^* = \sup_{s \le t} W_s$ and if $\lambda \ge 0$

$$\mathbf{P}(W_t \le \lambda, \ W_t^* \ge \lambda) = \mathbf{P}(W_t \ge \lambda, \ W_t^* \ge \lambda) = \mathbf{P}(W_t \ge \lambda).$$

Conclude that W_t^* and $|W_t|$ have the same probability law.

6. If $\lambda \geq \mu$ and $\lambda \geq 0$, prove that

$$\mathbf{P}(W_t \leq \mu, \ W_t^* \geq \lambda) = \mathbf{P}(W_t \geq 2\lambda - \mu, \ W_t^* \geq \lambda) = \mathbf{P}(W_t \geq 2\lambda - \mu),$$

and if $\lambda \leq \mu$ and $\lambda \geq 0$

$$\mathbf{P}(W_t \leq \mu, \ W_t^* \geq \lambda) = 2\mathbf{P}(W_t \geq \lambda) - \mathbf{P}(W_t \geq \mu).$$

7. Finally, check that the law of (W_t, W_t^*) is given by

$$1_{\{0 \leq y\}} 1_{\{x \leq y\}} \frac{2(2y - x)}{\sqrt{2\pi t^3}} \exp\left(-\frac{(2y - x)^2}{2t}\right) dx dy.$$

4

The Black-Scholes model

Black and Scholes (1973) tackled the problem of pricing and hedging a European option (call or put) on a non-dividend paying stock. Their method, which is based on similar ideas to those developed in discrete-time in Chapter 1 of this book, leads to some formulae frequently used by practitioners, despite the simplifying character of the model. In this chapter, we give an up-to-date presentation of their work. The case of the American option is investigated and some extensions of the model are exposed.

4.1 Description of the model

4.1.1 The behaviour of prices

The model suggested by Black and Scholes to describe the behaviour of prices is a continuous-time model with one risky asset (a share with price S_t at time t) and a riskless asset (with price S_t^0 at time t). We suppose the behaviour of S_t^0 to be encapsulated by the following (ordinary) differential equation:

$$dS_t^0 = rS_t^0 dt, \tag{4.1}$$

where r is a non-negative constant. Note that r is an instantaneous interest rate and should not be confused with the one-period rate in discrete-time models. We set $S_0^0 = 1$, so that $S_t^0 = e^{rt}$ for $t \geq 0$.

We assume that the behaviour of the stock price is determined by the following stochastic differential equation:

$$dS_t = S_t \left(\mu dt + \sigma dB_t \right), \tag{4.2}$$

where μ and σ are two constants and (B_t) is a standard Brownian motion.

The model is valid on the interval $[0, T]$ where T is the maturity of the option. As we saw (Chapter 3, Section 3.4.3), equation (4.2) has a closed-form solution

$$S_t = S_0 \exp \left(\mu t - \frac{\sigma^2}{2} t + \sigma B_t \right),$$

where S_0 is the spot price observed at time 0. One particular result from this model is that the law of S_t is lognormal (i.e. its logarithm follows a normal law).

More precisely, we see that the process (S_t) is a solution of an equation of type (4.2) if and only if the process $(\log(S_t))$ is a Brownian motion (not necessarily standard). According to Definition 3.2.1 of Chapter 3, the process (S_t) has the following properties:

- continuity of the sample paths;

- independence of the relative increments: if $u \le t$, S_t/S_u or (equivalently), the relative increment $(S_t - S_u)/S_u$ is independent of the σ-algebra $\sigma(S_v, v \le u)$;

- stationarity of the relative increments: if $u \le t$, the law of $(S_t - S_u)/S_u$ is identical to the law of $(S_{t-u} - S_0)/S_0$.

These three properties express in concrete terms the hypotheses of Black and Scholes on the behaviour of the share price.

4.1.2 Self-financing strategies

A strategy will be defined as a process $\phi = (\phi_t)_{0 \le t \le T} = ((H_t^0, H_t))$ with values in $\mathbb{R}^2$, adapted to the natural filtration $(\mathcal{F}_t)$ of the Brownian motion; the components H_t^0 and H_t are the quantities of riskless asset and risky asset respectively, held in the portfolio at time t. The value of the portfolio at time t is then given by

$$V_t(\phi) = H_t^0 S_t^0 + H_t S_t.$$

In the discrete-time models, we have characterised self-financing strategies by the equality: $V_{n+1}(\phi) - V_n(\phi) = \phi_{n+1}.(S_{n+1} - S_n)$ (see Chapter 1, Remark 1.1.1). This equality is extended to give the self-financing condition in the continuous-time case

$$dV_t(\phi) = H_t^0 dS_t^0 + H_t dS_t.$$

To give a meaning to this equality, we set the condition

$$\int_0^T |H_t^0| dt < +\infty \text{ a.s.} \quad \text{and} \quad \int_0^T H_t^2 dt < +\infty \text{ a.s.}$$

Then the integral

$$\int_0^T H_t^0 dS_t^0 = \int_0^T H_t^0 r e^{rt} dt$$

is well-defined, as is the stochastic integral

$$\int_0^T H_t dS_t = \int_0^T (H_t S_t \mu) \, dt + \int_0^T \sigma H_t S_t dB_t,$$

since the map $t \mapsto S_t$ is continuous, thus bounded on $[0, T]$ almost surely.

Definition 4.1.1 *A self-financing strategy is defined by a pair ϕ of adapted processes $(H_t^0)_{0 \le t \le T}$ and $(H_t)_{0 \le t \le T}$ satisfying:*

1. $\displaystyle\int_0^T |H_t^0|\,dt + \int_0^T H_t^2\,dt < +\infty$ a.s.

2. $\displaystyle H_t^0 S_t^0 + H_t S_t = H_0^0 S_0^0 + H_0 S_0 + \int_0^t H_u^0\,dS_u^0 + \int_0^t H_u\,dS_u$ a.s.

   for all $t \in [0, T]$.

We denote by $\tilde{S}_t = e^{-rt} S_t$ the discounted price of the risky asset. The following proposition is the counterpart of Proposition 1.1.2 of Chapter 1.

Proposition 4.1.2 *Let $\phi = ((H_t^0, H_t))_{0 \le t \le T}$ be an adapted process with values in $\mathbb{R}^2$, satisfying $\int_0^T |H_t^0|\,dt + \int_0^T H_t^2\,dt < +\infty$ a.s. We set: $V_t(\phi) = H_t^0 S_t^0 + H_t S_t$ and $\tilde{V}_t(\phi) = e^{-rt} V_t(\phi)$. Then, ϕ defines a self-financing strategy if and only if*

$$\tilde{V}_t(\phi) = V_0(\phi) + \int_0^t H_u\,d\tilde{S}_u \quad a.s. \tag{4.3}$$

for all $t \in [0, T]$.

Proof. Let us consider the self-financing strategy ϕ. From equality

$$d\tilde{V}_t(\phi) = -r\tilde{V}_t(\phi)dt + e^{-rt}dV_t(\phi)$$

which results from the differentiation of the product of the processes (e^{-rt}) and $(V_t(\phi))$ (the cross-variation term $d\langle e^{-r\cdot}, V.(\phi)\rangle_t$ is null), we deduce

$$
\begin{aligned}
d\tilde{V}_t(\phi) &= -re^{-rt}\left(H_t^0 e^{rt} + H_t S_t\right)dt + e^{-rt}H_t^0 d(e^{rt}) + e^{-rt}H_t dS_t \\
&= H_t\left(-re^{-rt}S_t dt + e^{-rt}dS_t\right) \\
&= H_t d\tilde{S}_t,
\end{aligned}
$$

which yields equality (4.3). The converse is justified similarly. $\qquad\square$

Remark 4.1.3 We have not imposed any condition of *predictability* on strategies unlike in Chapter 1. Actually, it is still possible to define a predictable process in continuous-time but, in the case of the filtration of a Brownian motion, it does not restrict the class of adapted processes significantly (because of the continuity of sample paths of Brownian motion).

In our study of complete discrete models, we had to consider at some stage a probability measure equivalent to the initial probability and under which discounted prices of assets are martingales. We were then able to design self-financing strategies replicating the option. The following section provides the tools which allow us to apply these methods in continuous time.

4.2 Change of probability. Representation of martingales

4.2.1 Equivalent probabilities

Let $(\Omega, \mathcal{A}, \mathbf{P})$ be a probability space. A probability measure $\mathbf{Q}$ on $(\Omega, \mathcal{A})$ is absolutely continuous relative to $\mathbf{P}$ if

$$\forall A \in \mathcal{A} \quad \mathbf{P}(A) = 0 \Rightarrow \mathbf{Q}(A) = 0.$$

Theorem 4.2.1 **Q** *is absolutely continuous relative to* **P** *if and only if there exists a non-negative random variable* Z *on* $(\Omega, \mathcal{A})$ *such that*

$$\forall A \in \mathcal{A} \quad \mathbf{Q}(A) = \int_A Z(\omega) d\mathbf{P}(\omega).$$

Z *is called* density *of* **Q** *relative to* **P** *and sometimes denoted* $d\mathbf{Q}/d\mathbf{P}$.

The sufficiency of the proposition is obvious, the converse is a version of the Radon-Nikodym theorem (cf. for example Dacunha-Castelle and Duflo (1986), Volume 1, or Williams (1991) Section 5.14).

The probabilities **P** and **Q** are *equivalent* if each one is absolutely continuous relative to the other. Note that if **Q** is absolutely continuous relative to **P**, with density Z, then **P** and **Q** are equivalent if and only if $\mathbf{P}(Z > 0) = 1$.

4.2.2 The Girsanov theorem

Let $\left(\Omega, \mathcal{F}, (\mathcal{F}_t)_{0 \leq t \leq T}, \mathbf{P}\right)$ be a probability space equipped with the natural filtration of a standard Brownian motion $(B_t)_{0 \leq t \leq T}$, indexed on the time interval $[0, T]$. The following theorem, which we admit, is known as the Girsanov theorem (cf. Karatzas and Shreve (1988), Dacunha-Castelle and Duflo (1986), Chapter 8).

Theorem 4.2.2 *Let* $(\theta_t)_{0 \leq t \leq T}$ *be an adapted process satisfying* $\int_0^T \theta_s^2 ds < \infty$ *a.s. and such that the process* $(L_t)_{0 \leq t \leq T}$ *defined by*

$$L_t = \exp\left(-\int_0^t \theta_s dB_s - \frac{1}{2}\int_0^t \theta_s^2 ds\right)$$

is a martingale. Then, under the probability $\mathbf{P}^{(L)}$ *with density* L_T *relative to* **P**, *the process* $(W_t)_{0 \leq t \leq T}$ *defined by* $W_t = B_t + \int_0^t \theta_s ds$, *is a standard Brownian motion.*

Remark 4.2.3 A sufficient condition for $(L_t)_{0 \leq t \leq T}$ to be a martingale is:

$$\mathbf{E}\left(\exp\left(\frac{1}{2}\int_0^T \theta_t^2 dt\right)\right) < \infty,$$

(see Karatzas and Shreve (1988), Dacunha-Castelle and Duflo (1986)). The proof of Girsanov theorem when (θ_t) is constant is the purpose of Exercise 19.

4.2.3 Representation of Brownian martingales

Let $(B_t)_{0 \leq t \leq T}$ be a standard Brownian motion built on a probability space $(\Omega, \mathcal{F}, \mathbf{P})$ and let $(\mathcal{F}_t)_{0 \leq t \leq T}$ be its *natural filtration*. Let us recall (see Chapter 3, Proposition 3.4.4) that if $(H_t)_{0 \leq t \leq T}$ is an adapted process such that $\mathbf{E}\left(\int_0^T H_t^2 dt\right) < \infty$, the process $\left(\int_0^t H_s dB_s\right)$ is a square-integrable martingale, null at 0. The following theorem shows that any Brownian martingale can be represented in terms of a stochastic integral.

Theorem 4.2.4 *Let $(M_t)_{0 \le t \le T}$ be a square-integrable martingale, with respect to the filtration $(\mathcal{F}_t)_{0 \le t \le T}$. There exists an adapted process $(H_t)_{0 \le t \le T}$ such that* $\mathbf{E}\left(\int_0^T H_s^2 ds\right) < +\infty$ *and*

$$\forall t \in [0,T] \quad M_t = M_0 + \int_0^t H_s dB_s \quad a.s. \tag{4.4}$$

Note that this representation only applies to martingales relative to the *natural* filtration of the Brownian motion (cf. Exercise 26).

From this theorem, it follows that if U is an $\mathcal{F}_T$-measurable, square-integrable random variable, it can be written as

$$U = \mathbf{E}(U) + \int_0^T H_s dB_s \quad a.s.,$$

where (H_t) is an adapted process such that $\mathbf{E}\left(\int_0^T H_s^2 ds\right) < +\infty$. To prove it, consider the martingale $M_t = \mathbf{E}(U | \mathcal{F}_t)$. It can also be shown (see, for example, Karatzas and Shreve (1988)) that if $(M_t)_{0 \le t \le T}$ is a martingale (not necessarily square-integrable) there is a representation similar to (4.4) with a process satisfying only $\int_0^T H_t^2 ds < \infty$, a.s. We will use this result in Chapter 6.

4.3 Pricing and hedging options in the Black-Scholes model

4.3.1 A probability under which $\left(\tilde{S}_t\right)$ is a martingale

We now consider the model introduced in Section 4.1. We will prove that there exists a probability equivalent to $\mathbf{P}$, under which the discounted share price $\tilde{S}_t = e^{-rt} S_t$ is a martingale. From the stochastic differential equation satisfied by (S_t), we have

$$\begin{aligned} d\tilde{S}_t &= -re^{-rt} S_t dt + e^{-rt} dS_t \\ &= \tilde{S}_t \left((\mu - r)dt + \sigma dB_t\right). \end{aligned}$$

Consequently, if we set $W_t = B_t + (\mu - r)t/\sigma$,

$$d\tilde{S}_t = \tilde{S}_t \sigma dW_t. \tag{4.5}$$

From Theorem 4.2.2, with $\theta_t = (\mu - r)/\sigma$, there exists a probability $\mathbf{P}^*$ equivalent to $\mathbf{P}$ under which $(W_t)_{0 \le t \le T}$ is a standard Brownian motion. We will admit that the definition of the stochastic integral is invariant by change of equivalent probability (cf. Exercise 25). Then, under the probability $\mathbf{P}^*$, we deduce from equality (4.5) that $(\tilde{S}_t)$ is a martingale and that

$$\tilde{S}_t = \tilde{S}_0 \exp(\sigma W_t - \sigma^2 t/2).$$

4.3.2 Pricing

In this section, we will focus on European options. A European option will be defined by a non-negative, $\mathcal{F}_T$-measurable, random variable h. Quite often, h can be written as $f(S_T)$, ($f(x) = (x - K)_+$ in the case of a call, $f(x) = (K - x)_+$ in the case of a put). As in the discrete-time setting, we will define the option value by a replication argument. For technical reasons, we will limit our study to the following admissible strategies:

Definition 4.3.1 *A strategy $\phi = \left(H_t^0, H_t\right)_{0 \leq t \leq T}$ is admissible if it is self-financing and if the discounted value $\tilde{V}_t(\phi) = H_t^0 + H_t \tilde{S}_t$ of the corresponding portfolio is, for all t, non-negative and such that $\sup_{t \in [0,T]} \tilde{V}_t$ is square-integrable under $\mathbf{P}^*$.*

An option is said to be replicable if its payoff at maturity is equal to the final value of an admissible strategy. It is clear that, for the option defined by h to be replicable, it is necessary that h should be square-integrable under $\mathbf{P}^*$. In the case of a call ($h = (S_T - K)_+$), this property indeed holds since $\mathbf{E}^*(S_T^2) < \infty$; note that in the case of a put, h is even bounded.

Theorem 4.3.2 *In the Black-Scholes model, any option defined by a non-negative, $\mathcal{F}_T$-measurable random variable h, which is square-integrable under the probability $\mathbf{P}^*$, is replicable and the value at time t of any replicating portfolio is given by*

$$V_t = \mathbf{E}^* \left(e^{-r(T-t)} h | \mathcal{F}_t\right).$$

Thus, the option value at time t can be naturally defined by the expression $\mathbf{E}^* \left(e^{-r(T-t)} h | \mathcal{F}_t\right)$.

Proof. First, let us assume that there is an admissible strategy (H^0, H), replicating the option. The value at time t of the portfolio (H_t^0, H_t) is given by

$$V_t = H_t^0 S_t^0 + H_t S_t,$$

and, by hypothesis, we have $V_T = h$. Let $\tilde{V}_t = V_t e^{-rt}$ be the discounted value

$$\tilde{V}_t = H_t^0 + H_t \tilde{S}_t.$$

Since the strategy is self-financing, we get from Proposition 4.1.2 and equality (4.5)

$$\begin{aligned} \tilde{V}_t &= V_0 + \int_0^t H_u d\tilde{S}_u \\ &= V_0 + \int_0^t H_u \sigma \tilde{S}_u dW_u. \end{aligned}$$

Under the probability $\mathbf{P}^*$, $\sup_{t \in [0,T]} \tilde{V}_t$ is square-integrable, by definition of admissible strategies. Furthermore, the preceding equality shows that the process $(\tilde{V}_t)$ is a stochastic integral relative to (W_t). It follows (cf. Chapter 3, Proposition 3.4.4 and Exercise 15) that $(\tilde{V}_t)$ is a square-integrable martingale under $\mathbf{P}^*$. Hence

$$\tilde{V}_t = \mathbf{E}^* \left(\tilde{V}_T | \mathcal{F}_t\right),$$

and consequently

$$V_t = \mathbf{E}^* \left(e^{-r(T-t)} h | \mathcal{F}_t \right). \tag{4.6}$$

So we have proved that if a portfolio (H^0, H) replicates the option defined by h, its value is given by equality (4.6). To complete the proof of the theorem, it remains to show that the option is indeed replicable, i.e. to find some processes (H_t^0) and (H_t) defining an admissible strategy, such that

$$H_t^0 S_t^0 + H_t S_t = \mathbf{E}^* \left(e^{-r(T-t)} h | \mathcal{F}_t \right).$$

Under the probability $\mathbf{P}^*$, the process defined by $M_t = \mathbf{E}^*(e^{-rT} h | \mathcal{F}_t)$ is a square-integrable martingale. The filtration $(\mathcal{F}_t)$, which is the natural filtration of (B_t), is also the natural filtration of (W_t) and, from the theorem of representation of Brownian martingales, there exists an adapted process $(K_t)_{0 \le t \le T}$ such that $\mathbf{E}^* \left(\int_0^T K_s^2 ds \right) < +\infty$ and

$$\forall t \in [0, T] \quad M_t = M_0 + \int_0^t K_s dW_s \quad \text{a.s.}$$

The strategy $\phi = (H^0, H)$, with $H_t = K_t/(\sigma \tilde{S}_t)$ and $H_t^0 = M_t - H_t \tilde{S}_t$, is then, from Proposition 4.1.2 and equality (4.5), a self-financing strategy; its value at time t is given by

$$V_t(\phi) = e^{rt} M_t = \mathbf{E}^* \left(e^{-r(T-t)} h | \mathcal{F}_t \right).$$

This expression clearly shows that $V_t(\phi)$ is a non-negative random variable, with $\sup_{0 \le t \le T} V_t(\phi)$ square-integrable under $\mathbf{P}^*$ and that $V_T(\phi) = h$. We have found an admissible strategy replicating h. $\square$

Remark 4.3.3 When the random variable h can be written as $h = f(S_T)$, we can express the option value V_t at time t as a function of t and S_t. We have indeed

$$
\begin{aligned}
V_t &= \mathbf{E}^* \left(e^{-r(T-t)} f(S_T) | \mathcal{F}_t \right) \\
&= \mathbf{E}^* \left(e^{-r(T-t)} f \left(S_t e^{r(T-t)} e^{\sigma(W_T - W_t) - (\sigma^2/2)(T-t)} \right) \Big| \mathcal{F}_t \right).
\end{aligned}
$$

The random variable S_t is $\mathcal{F}_t$-measurable and, under $\mathbf{P}^*$, $W_T - W_t$ is independent of $\mathcal{F}_t$. Therefore, from Proposition A.2.5 of the Appendix, we deduce

$$V_t = F(t, S_t),$$

where

$$F(t, x) = \mathbf{E}^* \left(e^{-r(T-t)} f \left(x e^{r(T-t)} e^{\sigma(W_T - W_t) - (\sigma^2/2)(T-t)} \right) \right). \tag{4.7}$$

Since, under $\mathbf{P}^*$, $W_T - W_t$ is a zero-mean normal variable with variance $T - t$

$$F(t, x) = e^{-r(T-t)} \int_{-\infty}^{+\infty} f \left(x e^{(r - \sigma^2/2)(T-t) + \sigma y \sqrt{T-t}} \right) \frac{e^{-y^2/2} dy}{\sqrt{2\pi}}.$$

F can be calculated explicitly for calls and puts. If we choose the case of the call, where $f(x) = (x - K)_+$, we have, from equality (4.7)

$$
\begin{aligned}
F(t, x) &= \mathbf{E}^* \left(e^{-r(T-t)} \left(xe^{(r-\sigma^2/2)(T-t)+\sigma(W_T - W_t)} - K \right)_+ \right) \\
&= \mathbf{E} \left(xe^{\sigma\sqrt{\theta}g - \sigma^2\theta/2} - Ke^{-r\theta} \right)_+
\end{aligned}
$$

where g is a standard Gaussian variable and $\theta = T - t$.

Let us set

$$
d_1 = \frac{\log(x/K) + (r + \sigma^2/2)\,\theta}{\sigma\sqrt{\theta}} \quad \text{and} \quad d_2 = d_1 - \sigma\sqrt{\theta}.
$$

Using these notations, we have

$$
\begin{aligned}
F(t, x) &= \mathbf{E} \left[\left(xe^{\sigma\sqrt{\theta}g - \sigma^2\theta/2} - Ke^{-r\theta} \right) 1_{\{g+d_2 \geq 0\}} \right] \\
&= \int_{-d_2}^{+\infty} \left(xe^{\sigma\sqrt{\theta}y - \sigma^2\theta/2} - Ke^{-r\theta} \right) \frac{e^{-y^2/2}}{\sqrt{2\pi}} dy \\
&= \int_{-\infty}^{d_2} \left(xe^{-\sigma\sqrt{\theta}y - \sigma^2\theta/2} - Ke^{-r\theta} \right) \frac{e^{-y^2/2}}{\sqrt{2\pi}} dy.
\end{aligned}
$$

Writing this expression as the difference of two integrals and in the first one using the change of variable $z = y + \sigma\sqrt{\theta}$, we obtain

$$
F(t, x) = xN(d_1) - Ke^{-r\theta}N(d_2), \tag{4.8}
$$

where

$$
N(d) = \frac{1}{\sqrt{2\pi}} \int_{-\infty}^{d} e^{-x^2/2} dx.
$$

Using identical notations and through similar calculations, the price of the put is equal to

$$
F(t, x) = Ke^{-r\theta}N(-d_2) - xN(-d_1). \tag{4.9}
$$

The reader will find efficient methods to compute $N(d)$ in Chapter 8.

Remark 4.3.4 One of the main features of the Black-Scholes model (and one of the reasons for its success) is the fact that the pricing formulae, as well as the hedging formulae we will give later, depend on only one non-observable parameter: σ, called 'volatility' by practitioners (the drift parameter μ disappears by change of probability). In practice, two methods are used to evaluate σ:

1. The historical method: in the present model, $\sigma^2 T$ is the variance of $\log(S_T)$ and the variables $\log(S_T/S_0)$, $\log(S_{2T}/S_T)$, ..., $\log(S_{NT}/S_{(N-1)T})$ are independent and identically distributed. Therefore, σ can be estimated by statistical means using the asset prices observed in the past (for example by calculating empirical variances; cf. Dacunha-Castelle and Duflo (1986), Chapter 5).

2. The 'implied' method: some options are quoted on organised markets; the price of options (calls and puts) being an increasing function of σ (cf. Exercise 21), we can associate an 'implied' volatility to each quoted option, by inversion of the Black-Scholes formula. Once the model is identified, it can be used to elaborate hedging schemes.

In those problems concerning volatility, one is soon confronted with the imperfections of the Black-Scholes model. Important differences between historical volatility and implied volatility are observed, the latter seeming to depend upon the strike price and the maturity. In spite of these incoherences, the model is considered as a reference by practitioners.

4.3.3 Hedging calls and puts

In the proof of Theorem 4.3.2, we referred to the theorem of representation of Brownian martingales to show the existence of a replicating portfolio. In practice, a theorem of existence is not satisfactory and it is essential to be able to build a real replicating portfolio to hedge an option.

When the option is defined by a random variable $h = f(S_T)$, we show that it is possible to find an explicit hedging portfolio. A replicating portfolio must have, at any time t, a discounted value equal to

$$\tilde{V}_t = e^{-rt}F(t, S_t),$$

where F is the function defined by equality (4.7). Under large hypothesis on f (and, in particular, in the case of calls and puts where we have the closed-form solutions of Remark 4.3.3), we see that the function F is of class C^∞ on $[0, T[\times\mathbb{R}$. If we set

$$\tilde{F}(t, x) = e^{-rt}F\left(t, xe^{rt}\right),$$

we have $\tilde{V}_t = \tilde{F}(t, \tilde{S}_t)$ and, for $t < T$, from the Itô formula

$$
\begin{aligned}
\tilde{F}\left(t, \tilde{S}_t\right) &= \tilde{F}\left(0, \tilde{S}_0\right) + \int_0^t \frac{\partial \tilde{F}}{\partial x}\left(u, \tilde{S}_u\right) d\tilde{S}_u \\
&+ \int_0^t \frac{\partial \tilde{F}}{\partial t}\left(u, \tilde{S}_u\right) du + \int_0^t \frac{1}{2}\frac{\partial^2 \tilde{F}}{\partial x^2}\left(u, \tilde{S}_u\right) d\langle \tilde{S}, \tilde{S} \rangle_u
\end{aligned}
$$

From equality $d\tilde{S}_t = \sigma \tilde{S}_t dW_t$, we deduce

$$d\langle \tilde{S}, \tilde{S} \rangle_u = \sigma^2 \tilde{S}_u^2 du,$$

so that $\tilde{F}\left(t, \tilde{S}_t\right)$ can be written as

$$\tilde{F}\left(t, \tilde{S}_t\right) = \tilde{F}\left(0, \tilde{S}_0\right) + \int_0^t \sigma\frac{\partial \tilde{F}}{\partial x}\left(u, \tilde{S}_u\right) \tilde{S}_u dW_u + \int_0^t K_u du.$$

Since $\tilde{F}(t, \tilde{S}_t)$ is a martingale under $\mathbf{P}^*$, the process K_u is necessarily null (cf.

Chapter 3, Exercise 16). Hence

$$\tilde{F}\left(t, \tilde{S}_t\right) = \tilde{F}\left(0, \tilde{S}_0\right) + \int_0^t \sigma \frac{\partial \tilde{F}}{\partial x}\left(u, \tilde{S}_u\right) \tilde{S}_u dW_u$$

$$= \tilde{F}\left(0, \tilde{S}_0\right) + \int_0^t \frac{\partial \tilde{F}}{\partial x}\left(u, \tilde{S}_u\right) d\tilde{S}_u.$$

The natural candidate for the hedging process H_t is then

$$H_t = \frac{\partial \tilde{F}}{\partial x}\left(t, \tilde{S}_t\right) = \frac{\partial F}{\partial x}\left(t, S_t\right).$$

If we set $H_t^0 = \tilde{F}\left(t, \tilde{S}_t\right) - H_t \tilde{S}_t$, the portfolio (H_t^0, H_t) is self-financing and its discounted value is indeed $\tilde{V}_t = \tilde{F}\left(t, \tilde{S}_t\right)$.

Remark 4.3.5 The preceding argument shows that it is not absolutely necessary to use the theorem of representation of Brownian martingales to deal with options of the form $f(S_T)$.

Remark 4.3.6 In the case of the call, we have, using the same notations as in Remark 4.3.3,

$$\frac{\partial F}{\partial x}(t, x) = N(d_1),$$

and in the case of the put

$$\frac{\partial F}{\partial x}(t, x) = -N(-d_1).$$

This is left as an exercise (the easiest way is to differentiate under the expectation). This quantity is often called the 'delta' of the option by practitioners. More generally, when the value at time t of a portfolio can be expressed as $\Psi(t, S_t)$, the quantity $(\partial \Psi / \partial x)(t, S_t)$, which measures the sensitivity of the portfolio with respect to the variations of the asset price at time t, is called the 'delta' of the portfolio. 'gamma' refers to the second-order derivative $(\partial^2 \Psi / \partial x^2)(t, S_t)$, 'theta' to the derivative with respect to time and 'vega' to the derivative of Ψ with respect to the volatility σ.

4.4 American options in the Black-Scholes model

4.4.1 Pricing American options

We have seen in Chapter 2 how the pricing of American options and the optimal stopping problem are related in discrete-time setting. The theory of optimal stopping in continuous-time is based on the same ideas as in discrete-time but is far more complex technically speaking. The approach we proposed in Section 1.3.3 of Chapter 1, based on an induction argument, cannot be used directly to price American options. Exercise 5 in Chapter 2 shows that, in a discrete model, it is possible to associate any American option to a hedging scheme with consumption.

Definition 4.4.1 *A trading strategy with consumption is defined as an adapted process* $\phi = ((H_t^0, H_t))_{0 \le t \le T}$, *with values in* $\mathbb{R}^2$, *satisfying the following properties:*

1. $\displaystyle\int_0^T |H_t^0| dt + \int_0^T H_t^2 dt < +\infty$ *a.s.*

2. $H_t^0 S_t^0 + H_t S_t = H_0^0 S_0^0 + H_0 S_0 + \displaystyle\int_0^t H_u^0 dS_u^0 + \int_0^t H_u dS_u - C_t$ *for all*

   $t \in [0, T]$, *where* $(C_t)_{0 \le t \le T}$ *is an adapted, continuous, non-decreasing process null at* $t = 0$; C_t *corresponds to the* cumulative consumption *up to time t.*

An American option is naturally defined by an adapted non-negative process $(h_t)_{0 \le t \le T}$. To simplify, we will only study processes of the form $h_t = \psi(S_t)$, where ψ is a continuous function from $\mathbb{R}^+$ to $\mathbb{R}^+$, satisfying: $\psi(x) \le A + Bx$, $\forall x \in \mathbb{R}^+$, for some non-negative constants A and B. For a call, we have: $\psi(x) = (x - K)_+$ and for a put: $\psi(x) = (K - x)_+$.

The trading strategy with consumption $\phi = ((H_t^0, H_t))_{0 \le t \le T}$ is said to hedge the American option defined by $h_t = \psi(S_t)$ if, setting $V_t(\phi) = H_t^0 S_t^0 + H_t S_t$, we have

$$\forall t \in [0, T] \quad V_t(\phi) \ge \psi(S_t) \text{ a.s.}$$

Denote by Φ^ψ the set of trading strategies with consumption hedging the American option defined by $h_t = \psi(S_t)$. If the writer of the option follows a strategy $\phi \in \Phi^\psi$, he or she possesses at any time t, a wealth at least equal to $\psi(S_t)$, which is precisely the payoff if the option is exercised at time t. The following theorem introduces the minimal value of a hedging scheme for an American option:

Theorem 4.4.2 *Let* u *be the map from* $[0, T] \times \mathbb{R}^+$ *to* $\mathbb{R}$ *defined by*

$$u(t, x) = \sup_{\tau \in \mathcal{T}_{t,T}} \mathbf{E}^* \left[e^{-r(\tau - t)} \psi \left(x \exp \left((r - (\sigma^2/2))(\tau - t) + \sigma(W_\tau - W_t) \right) \right) \right]$$

where $\mathcal{T}_{t,T}$ *represents the set of stopping times taking values in* $[t, T]$. *There exists a strategy* $\bar{\phi} \in \Phi^\psi$ *such that* $V_t(\bar{\phi}) = u(t, S_t)$, *for all* $t \in [0, T]$. *Moreover, for any strategy* $\phi \in \Phi^\psi$, *we have:* $V_t(\phi) \ge u(t, S_t)$, *for all* $t \in [0, T]$.

To overcome technical difficulties, we give only the outlines of the proof (see Karatzas and Shreve (1988) for details). First, we show that the process $(e^{-rt}u(t, S_t))$ is the Snell envelope of the process $(e^{-rt}\psi(S_t))$, i.e. the smallest supermartingale which bounds it from above under $\mathbf{P}^*$. As it can be proved that the discounted value of a trading strategy with consumption is a supermartingale under $\mathbf{P}^*$, we deduce the inequality: $V_t(\phi) \ge u(t, S_t)$, for any strategy $\phi \in \Phi^\psi$. To show the existence of a strategy $\bar{\phi}$ such that $V_t(\bar{\phi}) = u(t, S_t)$, we have to use a theorem of decomposition of supermartingales similar to Proposition 2.3.1 of Chapter 2 as well as a theorem of representation of Brownian martingales.

It is natural to consider $u(t, S_t)$ as a price for the American option at time t, since it is the minimal value of a strategy hedging the option.

Remark 4.4.3 Let τ be a stopping time taking values in $[0, T]$. The value at time 0 of an admissible strategy in the sense of Definition 4.3.1 with value $\psi(S_\tau)$ at time τ is given by $\mathbf{E}^* (e^{-r\tau}\psi(S_\tau))$, since the discounted value of any admissible strategy is a martingale under $\mathbf{P}^*$. Thus the quantity $u(0, S_0) = \sup_{\tau \in \mathcal{T}_{0,T}} \mathbf{E}^*(e^{-r\tau}\psi(S_\tau))$ is the initial wealth that hedges the whole range of possible exercises.

As in discrete models, we notice that the American call price (on a non-dividend paying stock) is equal to the European call price:

Proposition 4.4.4 *If in Theorem 4.4.2, ψ is given by $\psi(x) = (x - K)_+$, for all real x, then we have*

$$u(t, x) = F(t, x)$$

where F is the function defined by equation (4.8) corresponding to the European call price.

Proof. We assume here that $t = 0$ (the proof is the same for $t > 0$). Then it is sufficient to show that, for any stopping time τ,

$$\mathbf{E}^*(e^{-r\tau}(S_\tau - K)_+) \leq \mathbf{E}^*(e^{-rT}(S_T - K)_+) = \mathbf{E}^*(\tilde{S}_T - e^{-rT}K)_+.$$

On the other hand, we have

$$\mathbf{E}^* \left((\tilde{S}_T - e^{-rT}K)_+ | \mathcal{F}_\tau \right) \geq \mathbf{E}^* \left((\tilde{S}_T - e^{-rT}K) | \mathcal{F}_\tau \right) = \tilde{S}_\tau - e^{-rT}K$$

since $(\tilde{S}_t)$ is a martingale under $\mathbf{P}^*$. Hence

$$\mathbf{E}^* \left((\tilde{S}_T - e^{-rT}K)_+ | \mathcal{F}_\tau \right) \geq \tilde{S}_\tau - e^{-r\tau}K$$

since $r \geq 0$ and by non-negativity of the left-hand term,

$$\mathbf{E}^* \left((\tilde{S}_T - e^{-rT}K)_+ | \mathcal{F}_\tau \right) \geq \left(\tilde{S}_\tau - e^{-r\tau}K \right)_+.$$

We obtain the desired inequality by computing the expectation of both sides. $\quad\square$

4.4.2 Perpetual puts, critical price

In the case of the put, the American option price is not equal to the European one and there is no closed-form solution for the function u. One has to use numerical methods; we present some of them in Chapter 5. In this section we will only use the formula

$$u(t, x) = \sup_{\tau \in \mathcal{T}_{t,T}} \mathbf{E}^* \left(Ke^{-r(\tau - t)} - x \exp \left(-\sigma^2(\tau - t)/2 + \sigma(W_\tau - W_t) \right) \right)_+$$

(4.10)

to deduce some properties of the function u. To make our point clearer, we assume $t = 0$. In fact, it is always possible to come down to this case by replacing T with

$T - t$. Equation (4.10) becomes

$$u(0, x) = \sup_{\tau \in \mathcal{T}_{0,T}} \mathbf{E}^* \left(Ke^{-r\tau} - x \exp \left(\sigma W_\tau - (\sigma^2 \tau/2) \right) \right)_+ . \qquad (4.11)$$

Let us consider a probability space $(\Omega, \mathcal{F}, \mathbf{P})$, and let $(B_t)_{0 \leq t < \infty}$ be a standard Brownian motion defined on this space and $\mathbb{R}^+$. Then, we get

$$
\begin{aligned}
u(0, x) &= \sup_{\tau \in \mathcal{T}_{0,T}} \mathbf{E} \left(Ke^{-r\tau} - x \exp \left(\sigma B_\tau - (\sigma^2 \tau/2) \right) \right)_+ \\
&\leq \sup_{\tau \in \mathcal{T}_{0,\infty}} \mathbf{E} \left[\left(Ke^{-r\tau} - x \exp \left(\sigma B_\tau - (\sigma^2 \tau/2) \right) \right)_+ 1_{\{\tau < \infty\}} \right],
\end{aligned}
$$

$$(4.12)$$

noting $\mathcal{T}_{0,\infty}$ the set of all stopping times of the filtration of $(B_t)_{t \geq 0}$ and $\mathcal{T}_{0,T}$ the set of all elements of $\mathcal{T}_{0,\infty}$ with values in $[0, T]$. The right-hand term in inequality (4.12) can be interpreted naturally as the value of a 'perpetual' put (i.e. it can be exercised at any time). The following proposition gives an explicit expression for the upper bound in (4.12).

Proposition 4.4.5 *The function*

$$u^\infty(x) = \sup_{\tau \in \mathcal{T}_{0,\infty}} \mathbf{E} \left[\left(Ke^{-r\tau} - x \exp \left(\sigma B_\tau - (\sigma^2 \tau/2) \right) \right)_+ 1_{\{\tau < \infty\}} \right] \qquad (4.13)$$

is given by the formulae

$$u^\infty(x) = K - x \quad \text{for} \quad x \leq x^*$$

$$u^\infty(x) = (K - x^*) \left(\frac{x}{x^*} \right)^{-\gamma} \quad \text{for} \quad x > x^*$$

with $x^* = K\gamma/(1 + \gamma)$ and $\gamma = 2r/\sigma^2$.

Proof. From formula (4.13) we deduce that the function u^∞ is convex, decreasing on $[0, \infty[$ and satisfies: $u^\infty(x) \geq (K - x)_+$ and, for any $T > 0$, $u^\infty(x) \geq \mathbf{E}(Ke^{-rT} - x \exp(\sigma B_T - (\sigma^2 T/2)))_+$, which implies: $u^\infty(x) > 0$, for all $x \geq 0$. Now we note $x^* = \sup\{x \geq 0 | u^\infty(x) = K - x\}$. From the properties of u^∞ we have just stated, it follows that

$$\forall x \leq x^* \quad u^\infty(x) = K - x \quad \text{and} \quad \forall x > x^* \quad u^\infty(x) > (K - x)_+. \qquad (4.14)$$

On the other hand, the Snell envelope theory in continuous time (cf. El Karoui (1981), Kushner (1977), as well as Chapter 5) enables us to show

$$u^\infty(x) = \mathbf{E} \left[\left(Ke^{-r\tau_x} - x \exp \left(\sigma B_{\tau_x} - (\sigma^2 \tau_x/2) \right) \right)_+ 1_{\{\tau_x < \infty\}} \right]$$

where τ_x is the stopping time defined by $\tau_x = \inf\{t \geq 0 | e^{-rt} u^\infty(X_t^x) = e^{-rt}(K - X_t^x)_+\}$ (with $\inf \emptyset = \infty$), the process (X_t^x) being defined by: $X_t^x = x \exp \left((r - \sigma^2/2)t + \sigma B_t \right)$. The stopping time τ_x is therefore an optimal stopping time (note the analogy with the results in Chapter 2).

It follows from (4.14) that

$$\tau_x = \inf\{t \geq 0 | X_t^x \leq x^*\} = \inf \left\{ t \geq 0 | (r - \sigma^2/2)t + \sigma B_t \leq \log(x^*/x) \right\}.$$

Introduce, for any $z \in \mathbb{R}_+$, the stopping time $\tau_{x,z}$ defined by

$$\tau_{x,z} = \inf\{t \geq 0 | X_t^x \leq z\}.$$

With these notations, the optimal stopping time is given by $\tau_x = \tau_{x,x^*}$. We fix x and we note ϕ the function of z defined by

$$\phi(z) = \mathbf{E}\left(e^{-r\tau_{x,z}} 1_{\{\tau_{x,z} < \infty\}} \left(K - X_{\tau_{x,z}}^x\right)_+\right).$$

Since τ_{x,x^*} is optimal, the function ϕ attains its maximum at point $z = x^*$. We are going to calculate ϕ explicitly, then we will maximise it to determine x^* and $u^\infty(x) = \phi(x^*)$.

If $z > x$, it is obvious that $\tau_{x,z} = 0$ and $\phi(z) = (K - x)_+$. If $z \leq x$, we have, by the continuity of the paths of $(X_t^x)_{t \geq 0}$,

$$\tau_{x,z} = \inf\{t \geq 0 | X_t^x = z\}$$

and consequently

$$
\begin{aligned}
\phi(z) &= (K - z)_+ \mathbf{E}\left(e^{-r\tau_{x,z}} 1_{\{\tau_{x,z} < \infty\}}\right) \\
&= (K - z)_+ \mathbf{E}\left(e^{-r\tau_{x,z}}\right)
\end{aligned}
$$

with, by convention, $e^{-r\infty} = 0$. Using the expression of X_t^x in terms of B_t, we see that, for $z \leq x$,

$$
\begin{aligned}
\tau_{x,z} &= \inf\left\{t \geq 0 \left| \left(r - \frac{\sigma^2}{2}\right) t + \sigma B_t = \log(z/x)\right.\right\} \\
&= \inf\left\{t \geq 0 \left| \mu t + B_t = \frac{1}{\sigma} \log(z/x)\right.\right\},
\end{aligned}
$$

with $\mu = r/\sigma - \sigma/2$. Thus, if we note, for any real b,

$$T_b = \inf\{t \geq 0 | \mu t + B_t = b\},$$

we get

$$
\phi(z) = \begin{cases}
(K - x)_+ & \text{if } z > x \\
(K - z)\mathbf{E}\left(\exp\left(-rT_{\log(z/x)/\sigma}\right)\right) & \text{if } z \in [0, x] \cap [0, K] \\
0 & \text{if } z \in [0, x] \cap [K, +\infty[.
\end{cases}
$$

The maximum of ϕ is attained on the interval $[0, x] \cap [0, K]$. Using the following formula (proved in Exercise 24)

$$\mathbf{E}\left(e^{-\alpha T_b}\right) = \exp\left(\mu b - |b|\sqrt{\mu^2 + 2\alpha}\right),$$

it can be seen that

$$\forall z \in [0, x] \cap [0, K] \quad \phi(z) = (K - z)\left(\frac{z}{x}\right)^\gamma,$$

where $\gamma = 2r/\sigma^2$. The derivative of this function is given by

$$\phi'(z) = \frac{z^{\gamma-1}}{x^\gamma}\left(K\gamma - (\gamma+1)z\right).$$

It results that if $x \leq K\gamma/(\gamma+1)$, $\max_z \phi(z) = \phi(x) = K-x$ and if $x > K\gamma/(\gamma+1)$, $\max_z \phi(z) = \phi(K\gamma/(\gamma+1))$, and we recognise the required expressions. $\square$

Remark 4.4.6 Let us come back to the American put with finite maturity T. Following the same arguments as in the beginning of the proof of Proposition 4.4.5, we see that, for any $t \in [0, T[$, there exists a real $s(t)$ satisfying

$$\forall x \leq s(t) \quad u(t, x) = K - x \quad \text{and} \quad \forall x > s(t) \quad u(t, x) > (K - x)_+. \quad (4.15)$$

Taking inequality (4.12) into account, we obtain $s(t) \geq x^*$, for all $t \in [0, T[$. The real number $s(t)$ is interpreted as the 'critical price' at time t: if the price of the underlying asset at time t is less than $s(t)$, the buyer of the option should exercise his or her option immediately; in the opposite case, he should keep it.

Notes: The presentation we have used, based on Girsanov theorem, is inspired by Harrison and Pliska (1981) (also refer to Bensoussan (1984) and Section 5.8 in Karatzas and Shreve (1988)). The initial approach of Black-Scholes (1973) and Merton (1973) consisted in deriving a partial differential equation satisfied by the the call price as a function of time and spot price. It is based on an arbitrage argument and the Itô formula. For more information on statistical estimation of the models' parameters, the reader should refer to Dacunha-Castelle and Duflo (1986) and Dacunha-Castelle and Duflo (1986) and to the references in these books.

4.5 Exercises

Exercise 19 The objective of this exercise is to prove the Girsanov theorem 4.2.2 in the special case where the process (θ_t) is constant. Let $(B_t)_{0 \leq t \leq T}$ be a standard Brownian motion with respect to the filtration $(\mathcal{F}_t)_{0 \leq t \leq T}$ and let μ be a real-valued number. We set, for $0 \leq t \leq T$, $L_t = \exp\left(-\mu B_t - (\mu^2/2)t\right)$.

1. Show that $(L_t)_{0 \leq t \leq T}$ is a martingale relative to the filtration $(\mathcal{F}_t)$ and that $\mathbf{E}(L_t) = 1$, for all $t \in [0, T]$.

2. We note $\mathbf{P}^{(L_t)}$ the density probability L_t with respect to the initial probability $\mathbf{P}$. Show that the probabilities $\mathbf{P}^{(L_T)}$ and $\mathbf{P}^{(L_t)}$ coincide on the σ-algebra $\mathcal{F}_t$.

3. Let Z be an $\mathcal{F}_T$-measurable, bounded random variable. Show that the conditional expectation of Z, under the probability $\mathbf{P}^{(L_T)}$, given $\mathcal{F}_t$, is

$$\mathbf{E}^{(L_T)}(Z|\mathcal{F}_t) = \frac{\mathbf{E}\left(ZL_T|\mathcal{F}_t\right)}{L_t}.$$

4. We set $W_t = \mu t + B_t$, for all $t \in [0, T]$. Show that for all real-valued u and for all s and t in $\in [0, T]$, with $s \leq t$, we have

$$\mathbf{E}^{(L_T)} \left(e^{iu(W_t - W_s)} \middle| \mathcal{F}_s \right) = e^{-u^2(t-s)/2}.$$

Conclude using Proposition A.2.2 of the Appendix.

Exercise 20 Show that the portfolio replicating a European option in the Black-Scholes model is unique (in a sense to be specified).

Exercise 21 We consider an option described by $h = f(S_T)$ and we note F the function of time and spot corresponding to the option price (cf. equation (4.7)).

1. Show that if f is non-decreasing (resp. non-increasing), $F(t, x)$ is a non-decreasing function (resp. non-increasing) of x.

2. We assume that f is convex. Show that $F(t, x)$ is a convex function of x, a decreasing function of t if $r = 0$ and a non-decreasing function of σ in any case. (Hint: first consider equation (4.7) and make use of Jensen's inequality: $\Phi(\mathbf{E}(X)) \leq \mathbf{E}(\Phi(X))$, where Φ is a convex function and X is a random variable such that X and $\Phi(X)$ are integrable.)

3. We note F_c (resp. F_p) the function F obtained when $f(x) = (x - K)_+$ (resp. $f(x) = (K - x)_+$). Prove that $F_c(t, .)$ and $F_p(t, .)$ are non-negative for $t < T$. Study the functions $F_c(t, .)$ and $F_p(t, .)$ in the neighbourhood of 0 and $+\infty$.

Exercise 22 Calculate under the initial probability $\mathbf{P}$, the probability that a European call is exercised.

Exercise 23 Justify formulae (4.8) and (4.9) and calculate for a call and a put the delta, the gamma, the theta and the vega (cf. Remark 4.3.6).

Exercise 24 Let $(B_t)_{t \geq 0}$ be a standard Brownian motion. For any real-valued μ and b, we set

$$T_b^\mu = \inf\{t \geq 0 \mid \mu t + B_t = b\}$$

with the convention: $\inf \emptyset = \infty$.

1. Use the Girsanov theorem to show the following equality:

$$\forall \alpha, t > 0 \quad \mathbf{E}\left(e^{-\alpha(T_b^\mu \wedge t)}\right) = \mathbf{E}\left(e^{-\alpha(T_b^0 \wedge t)} \exp\left(\mu B_{T_b^0 \wedge t} - \frac{\mu^2}{2} T_b^0 \wedge t\right)\right).$$

2. Prove the inequality

$$\forall \alpha, t > 0 \quad \mathbf{E}\left(e^{-\alpha(T_b^0 \wedge t)} \exp\left(\mu B_{T_b^0 \wedge t} - \frac{\mu^2}{2} T_b^0 \wedge t\right) 1_{\{t < T_b^0\}}\right) \leq e^{-\alpha t}.$$

3. Deduce from above and Proposition 3.3.6 that

$$\forall \alpha > 0 \quad \mathbf{E}\left(e^{-\alpha T_b^\mu} 1_{\{T_b^\mu < \infty\}}\right) = \exp\left(\mu b - |b|\sqrt{2\alpha + \mu^2}\right).$$

4. Calculate $\mathbf{P}\left(T_b^\mu < \infty\right)$.

Exercise 25

1. Let $\mathbf{P}$ and $\mathbf{Q}$ be two equivalent probabilities on a measurable space $(\Omega, \mathcal{A})$. Show that if a sequence (X_n) of random variables converges in probability under $\mathbf{P}$, it converges in probability under $\mathbf{Q}$ to the same limit.

2. Notations and hypothesis are those of Theorem 4.2.2. Let $(H_t)_{0 \leq t \leq T}$ be an adapted process such that $\int_0^T H_s^2 ds < \infty$ $\mathbf{P}$-a.s. The stochastic integral of (H_t) relative to B_t is well-defined under the probability $\mathbf{P}$. We set

$$X_t = \int_0^t H_s dB_s + \int_0^t H_s \theta_s ds.$$

Since $\mathbf{P}^{(L)}$ and $\mathbf{P}$ are equivalent, we have $\int_0^T H_s^2 ds < \infty$ $\mathbf{P}^{(L)}$-a.s. and we can define, under $\mathbf{P}^{(L)}$, the process

$$Y_t = \int_0^t H_s dW_s.$$

The question is to prove the equality of the two processes X and Y. To do so, it is advised to consider first the case of simple processes; and to use the fact that if $(H_t)_{0 \leq t \leq T}$ is an adapted process satisfying $\int_0^T H_s^2 ds < \infty$ a.s., there is a sequence (H^n) of elementary processes such that $\int_0^T (H_s - H_s^n)^2 ds$ converges to 0 in probability.

Exercise 26 Let $(B_t)_{0 \leq t \leq 1}$ be a standard Brownian motion defined on the time interval $[0, 1]$. We note $(\mathcal{F}_t)_{0 \leq t \leq 1}$ its natural filtration and we consider τ an exponentially distributed random variable with parameter λ, independent of $\mathcal{F}_1$. For $t \in [0, 1]$, we note $\mathcal{G}_t$ the σ-algebra generated by $\mathcal{F}_t$ and the random variable $\tau \wedge t$.

1. Show that $(\mathcal{G}_t)_{0 \leq t \leq 1}$ is a filtration and that $(B_t)_{0 \leq t \leq 1}$ is a Brownian motion with respect to $(\mathcal{G}_t)$.

2. For $t \in [0, 1]$, we set $M_t = \mathbf{E}\left(1_{\{\tau > 1\}} | \mathcal{G}_t\right)$. Show that M_t is equal to $e^{-\lambda(1-t)} 1_{\{\tau > t\}}$ a.s. The following property can be used: if $\mathcal{B}_1$ and $\mathcal{B}_2$ are two sub-σ-algebras and X a non-negative random variable such that the σ-algebra generated by $\mathcal{B}_2$ and X are independent of the σ-algebra $\mathcal{B}_1$, then $\mathbf{E}\left(X | \mathcal{B}_1 \vee \mathcal{B}_2\right) = \mathbf{E}\left(X | \mathcal{B}_2\right)$, where $\mathcal{B}_1 \vee \mathcal{B}_2$ represents the σ-algebra generated by $\mathcal{B}_1$ and $\mathcal{B}_2$.

3. Show that there exists no path-continuous process (X_t) such that for all $t \in [0, 1]$, $\mathbf{P}(M_t = X_t) = 1$ (remark that we would necessarily have

$$\mathbf{P}\left(\forall t \in [0, 1] \ M_t = X_t\right) = 1).$$

Deduce that the martingale (M_t) cannot be represented as a stochastic integral with respect to (B_t).

Exercise 27 The reader may use the results of Exercise 18 of Chapter 3. Let $(W_t)_{t \geq 0}$ be an $\mathcal{F}_t$-Brownian motion.

1. Prove that if $\mu \leq \lambda$ and if $N(d) = \int_{-\infty}^{d} \exp(-x^2/2)(dx/(\sqrt{2\pi}))$

$$\mathbf{E}\left(e^{\alpha W_T}1_{\{W_T \leq \mu,\, \sup_{s \leq T} W_s \geq \lambda\}}\right) = \exp\left(\frac{\alpha^2 T}{2} + 2\alpha\lambda\right) N\left(\frac{\mu - 2\lambda - \alpha T}{\sqrt{T}}\right)$$

Deduce that if $\lambda \leq \mu$

$$\mathbf{E}\left(e^{\alpha W_T}1_{\{W_T \geq \mu,\, \inf_{s \leq T} W_s \leq \lambda\}}\right) = \exp\left(\frac{\alpha^2 T}{2} + 2\alpha\lambda\right) N\left(\frac{2\lambda - \mu + \alpha T}{\sqrt{T}}\right).$$

2. Let $H \leq K$; we are looking for an analytic formula for

$$C = \mathbf{E}\left(e^{-rT}(X_T - K)_+ 1_{\{\inf_{s \leq T} X_s \geq H\}}\right),$$

where $X_t = x \exp\left((r - \sigma^2/2)\, t + \sigma W_t\right)$. Give a financial interpretation to this value and give an expression for the probability $\tilde{P}$ that makes $\tilde{W}_t = (r/\sigma - \sigma/2)\, t + W_t$ a standard Brownian motion.

3. Write C as the expectation under $\tilde{P}$ of a random variable function only of $\tilde{W}_T$ and $\sup_{0 \leq s \leq T} \tilde{W}_s$.

4. Deduce an analytic formula for C.

Problem 1 Black-Scholes model with time-dependent parameters We consider once again the Black-Scholes model, assuming that the asset prices are described by the following equations (we keep the same notations as in this chapter)

$$\begin{cases} dS_t^0 &= r(t)S_t^0 dt \\ dS_t &= S_t(\mu(t)dt + \sigma(t)dB_t) \end{cases}$$

where $r(t)$, $\mu(t)$, $\sigma(t)$ are *deterministic* functions of time, continuous on $[0, T]$. Furthermore, we assume that $\inf_{t \in [0,T]} \sigma(t) > 0$:

1. Prove that

$$S_t = S_0 \exp\left(\int_0^t \mu(s)ds + \int_0^t \sigma(s)dB_s - \frac{1}{2}\int_0^t \sigma^2(s)ds\right).$$

You may consider the process

$$Z_t = S_t \exp - \left(\int_0^t \mu(s)ds + \int_0^t \sigma(s)dB_s - \frac{1}{2}\int_0^t \sigma^2(s)ds\right).$$

2.

   (a) Let (X_n) be a sequence of real-valued, zero-mean normal random variables converging to X in mean-square. Show that X is a normal random variable.

   (b) By approximating σ by simple functions, show that $\int_0^t \sigma(s)dB_s$ is a normal random variable and calculate its variance.

3. Prove that there exists a probability $\mathbf{P}^*$ equivalent to $\mathbf{P}$, under which the discounted stock price is a martingale. Give its density with respect to $\mathbf{P}$.

4. In the remainder, we will tackle the problem of pricing and hedging a call with maturity T and strike price K.

   (a) Let (H_t^0, H_t^1) be a self-financing strategy, with value V_t at time t. Show that if (V_t/S_t^0) is a martingale under $\mathbf{P}^*$ and if $V_T = (S_T - K)_+$, then

   $$\forall t \in [0, T] \quad V_t = F(t, S_t),$$

   where F is the function defined by

   $$F(t, x) = \mathbf{E}^* \left(x \, \exp \left(\int_t^T \sigma(s)dW_s - \frac{1}{2} \int_t^T \sigma^2(s)ds \right) - Ke^{-\int_t^T r(s)ds} \right)_+$$

   and (W_t) is a standard Brownian motion under $\mathbf{P}^*$.

   (b) Give an expression for the function F and compare it to the Black-Scholes formula.

   (c) Construct a hedging strategy for the call (find H_t^0 and H_t; check the self-financing condition).

Problem 2 Garman-Kohlhagen model

The Garman-Kohlhagen model (1983) is the most commonly used model to price and hedge foreign-exchange options. It derives directly from the Black-Scholes model. To clarify, we shall concentrate on 'dollar-franc' options. For example, a European *call* on the dollar, with maturity T and strike price K, is the right to buy, at time T, one dollar for K francs.

We will note S_t the price of the dollar at time t, i.e. the number of francs per dollar. The behaviour of S_t through time is modelled by the following stochastic differential equation:

$$\frac{dS_t}{S_t} = \mu dt + \sigma dW_t,$$

where $(W_t)_{t \in [0,T]}$ is a standard Brownian motion on a probability space $(\Omega, \mathcal{F}, \mathbf{P})$, μ and σ are real-valued, with $\sigma > 0$. We note $(\mathcal{F}_t)_{t \in [0,T]}$ the filtration generated by $(W_t)_{t \in [0,T]}$ and assume that $\mathcal{F}_t$ represents the accumulated information up to time t.

I

1. Express S_t as a function of S_0, t and W_t. Calculate the expectation of S_t.

2. Show that if $\mu > 0$, the process $(S_t)_{t \in [0,T]}$ is a submartingale.

3. Let $U_t = 1/S_t$ be the exchange rate of the franc against the dollar. Show that U_t satisfies the following stochastic differential equation

   $$\frac{dU_t}{U_t} = (\sigma^2 - \mu)dt - \sigma dW_t.$$

Deduce that if $0 < \mu < \sigma^2$, both processes $(S_t)_{t \in [0,T]}$ and $(U_t)_{t \in [0,T]}$ are submartingales. In what sense does it seem to be paradoxical?

II

We would like to price and hedge a European call on one dollar, with maturity T and strike price K, using a Black-Scholes-type method. From his premium, which represents his initial wealth, the writer of the option elaborates a strategy, defining at any time t a portfolio made of H_t^0 francs and H_t dollars, in order to create, at time T, a wealth equal to $(S_T - K)_+$ (in francs).

At time t, the value in francs of a portfolio made of H_t^0 francs and H_t dollars is obviously

$$V_t = H_t^0 + H_t S_t. \tag{4.16}$$

We suppose that French francs are invested or borrowed at the *domestic* rate r_0 and US dollars are invested or borrowed at the *foreign* rate r_1. A self-financing strategy will thus be defined by an adapted process $((H_t^0, H_t))_{t \in [0,T]}$, such that

$$dV_t = r_0 H_t^0 dt + r_1 H_t S_t dt + H_t dS_t \tag{4.17}$$

where V_t is defined by equation (4.16).

1. Which integrability conditions must be imposed on the processes (H_t^0) and (H_t) so that the differential equality (4.17) makes sense?

2. Let $\tilde{V}_t = e^{-r_0 t} V_t$ be the discounted value of the (self-financing) portfolio (H_t^0, H_t). Prove the equality

$$d\tilde{V}_t = H_t e^{-r_0 t} S_t (\mu + r_1 - r_0) dt + H_t e^{-r_0 t} S_t \sigma dW_t.$$

3.

   (a) Show that there exists a probability $\bar{\mathbf{P}}$, equivalent to $\mathbf{P}$, under which the process

   $$\bar{W}_t = \frac{\mu + r_1 - r_0}{\sigma} t + W_t$$

   is a standard Brownian motion.

   (b) A self-financing strategy is said to be *admissible* if its discounted value $\tilde{V}_t$ is non-negative for all t and if $\sup_{t \in [0,T]} \left(\tilde{V}_t \right)$ is square-integrable under $\bar{\mathbf{P}}$. Show that the discounted value of an admissible strategy is a martingale under $\bar{\mathbf{P}}$.

4. Show that if an admissible strategy replicates the call, in other words it is worth $V_T = (S_T - K)_+$ at time T, then for any $t \leq T$ the value of the strategy at time t is given by

$$V_t = F(t, S_t)$$

where

$$\begin{aligned} F(t,x) &= \bar{\mathbf{E}}\Big(x \exp\left(-(r_1 + (\sigma^2/2))(T - t) + \sigma(\bar{W}_T - \bar{W}_t)\right) \\ &\quad - K e^{-r_0(T-t)}\Big)_+ \end{aligned}$$

(The symbol $\bar{E}$ stands for the expectation under the probability $\bar{P}$.)

5. Show (through detailed calculation) that

$$F(t,x) = e^{-r_1(T-t)}xN(d_1) - Ke^{-r_0(T-t)}N(d_2)$$

where N is the distribution function of the standard normal law, and

$$d_1 = \frac{\log(x/K) + (r_0 - r_1 + (\sigma^2/2))(T-t)}{\sigma\sqrt{T-t}}$$

$$d_2 = \frac{\log(x/K) + (r_0 - r_1 - (\sigma^2/2))(T-t)}{\sigma\sqrt{T-t}}$$

6. The next step is to show that the option is effectively replicable.

   (a) We set $\bar{S}_t = e^{(r_1-r_0)t}S_t$. Derive the equality

   $$d\bar{S}_t = \sigma\bar{S}_t d\bar{W}_t.$$

   (b) Let $\bar{F}$ be the function defined by $\bar{F}(t,x) = e^{-r_0 t}F(t, xe^{(r_0-r_1)t})$ (F is the function defined in Question 4). We set $C_t = F(t, S_t)$ and $\tilde{C}_t = e^{-r_0 t}C_t = \bar{F}(t, \bar{S}_t)$. Derive the equality

   $$d\tilde{C}_t = \frac{\partial F}{\partial x}(t, S_t)\sigma e^{-r_0 t}S_t d\bar{W}_t.$$

   (c) Deduce that the call is replicable and give an explicit expression for the replicating portfolio $((H_t^0, H_t))$.

7. Write down a put-call parity relationship, similar to the relationship we gave for stocks, and give an example of arbitrage opportunity when this relationship does not hold.

Problem 3 Option to exchange one asset for another

We consider a financial market in which there are two risky assets with respective prices S_t^1 and S_t^2 at time t and a riskless asset with price $S_t^0 = e^{rt}$ at time t. The dynamics of the prices S_t^1 and S_t^2 over time are modelled by the following stochastic differential equations

$$\begin{cases} dS_t^1 = S_t^1\left(\mu_1 dt + \sigma_1 dB_t^1\right) \\ dS_t^2 = S_t^2\left(\mu_2 dt + \sigma_2 dB_t^2\right) \end{cases}$$

where $(B_t^1)_{t\in[0,T]}$ and $(B_t^2)_{t\in[0,T]}$ are two independent standard Brownian motions defined on a probability space $(\Omega, \mathcal{F}, \mathbf{P})$; μ_1, μ_2, σ_1 and σ_2 are real numbers, with $\sigma_1 > 0$ and $\sigma_2 > 0$. We note $\mathcal{F}_t$ the σ-algebra generated by the random variables B_s^1 and B_s^2 for $s \le t$. Then the processes $(B_t^1)_{t\in[0,T]}$ and $(B_t^2)_{t\in[0,T]}$ are $(\mathcal{F}_t)$-Brownian motions and, for $t \ge s$, the vector $(B_t^1 - B_s^1, B_t^2 - B_s^2)$ is independent of $\mathcal{F}_s$.

I

We study the pricing and hedging of an option giving the right to exchange one of the risky assets for the other at time T.

1. We set $\theta_1 = (\mu_1 - r)/\sigma_1$ and $\theta_2 = (\mu_2 - r)/\sigma_2$. Show that the process defined by

$$M_t = \exp\left(-\theta_1 B_t^1 - \theta_2 B_t^2 - \frac{1}{2}(\theta_1^2 + \theta_2^2)t\right),$$

   is a martingale with respect to the filtration $(\mathcal{F}_t)_{t \in [0,T]}$.

2. Let $\tilde{\mathbf{P}}$ be the probability with density M_T with respect to $\mathbf{P}$. We introduce the processes W^1 and W^2 defined by $W_t^1 = B_t^1 + \theta_1 t$ and $W_t^2 = B_t^2 + \theta_2 t$. Derive, under the probability $\tilde{\mathbf{P}}$, the joint characteristic function of (W_t^1, W_t^2). Deduce that, for any $t \in [0,T]$, the random variables W_t^1 and W_t^2 are independent normal random variables with zero-mean and variance t under $\tilde{\mathbf{P}}$.

   In the remainder, we will admit that, under the probability $\tilde{\mathbf{P}}$, the processes $(W_t^1)_{0 \le t \le T}$ and $(W_t^2)_{0 \le t \le T}$ are $(\mathcal{F}_t)$-independent standard Brownian motions and that, for $t \ge s$, the vector $(W_t^1 - W_s^1, W_t^2 - W_s^2)$ is independent of $\mathcal{F}_s$.

3. Write $\tilde{S}_t^1$ and $\tilde{S}_t^2$ as functions of S_0^1, S_0^2, W_t^1 and W_t^2 and show that, under $\tilde{\mathbf{P}}$, the discounted prices $\tilde{S}_t^1 = e^{-rt}S_t^1$ and $\tilde{S}_t^2 = e^{-rt}S_t^2$ are martingales.

We want to price and hedge a European option, with maturity T, giving to the holder the right to exchange one unit of the asset 2 for one unit of the asset 1. To do so, we use the same method as in the Black-Scholes model. From his initial wealth, the premium, the writer of the option builds a strategy, defining at any time t a portfolio made of H_t^0 units of the riskless asset and H_t^1 and H_t^2 units of the assets 1 and 2 respectively, in order to generate, at time T, a wealth equal to $(S_T^1 - S_T^2)_+$. A trading strategy will be defined by the three adapted processes H^0, H^1 and H^2.

II

1. Define precisely the self-financing strategies and prove that, if $\tilde{V}_t = e^{-rt}V_t$ is the discounted value of a self-financing strategy, we have

$$d\tilde{V}_t = H_t^1 e^{-rt}S_t^1 \sigma_1 dW_t^1 + H_t^2 e^{-rt}S_t^2 \sigma_2 dW_t^2.$$

2. Show that if the processes $(H_t^1)_{0 \le t \le T}$ and $(H_t^2)_{0 \le t \le T}$ of a self-financing strategy are uniformly bounded (which means that: $\exists C > 0, \forall(t,\omega) \in [0,T] \times \Omega, |H_t^i(\omega)| \le C$, for $i = 1, 2$), then the discounted value of the strategy is a martingale under $\tilde{\mathbf{P}}$.

3. Prove that if a self-financing strategy satisfies the hypothesis of the previous question and has a terminal value equal to $V_T = (S_T^1 - S_T^2)_+$ then its value at any time $t < T$ is given by

$$V_t = F(t, S_t^1, S_t^2), \tag{4.18}$$

where the function F is defined by

$$F(t, x_1, x_2) = \tilde{\mathbf{E}} \left(x_1 e^{\sigma_1(W_T^1 - W_t^1) - \frac{\sigma_1^2}{2}(T-t)} - x_2 e^{\sigma_2(W_T^2 - W_t^2) - \frac{\sigma_2^2}{2}(T-t)} \right)_+ ,$$
(4.19)

the symbol $\tilde{\mathbf{E}}$ representing the expectation under $\tilde{\mathbf{P}}$. The existence of a strategy having this value will be proved later on. We will consider in the remainder that the value of the option $(S_T^1 - S_T^2)_+$ at time t is given by $F(t, S_t^1, S_t^2)$.

4. Find a parity relationship between the value of the option with payoff $(S_T^1 - S_T^2)_+$ and the symmetrical option with payoff $(S_T^2 - S_T^1)_+$, similar to the put-call parity relationship previously seen and give an example of arbitrage opportunity when this relationship does not hold.

III

The objective of this section is to find an explicit expression for the function F defined by (4.19) and to establish a strategy replicating the option.

1. Let g_1 and g_2 be two independent standard normal random variables and let λ be a real number.

   (a) Show that under the probability $\mathbf{P}^{(\lambda)}$, with density with respect to $\mathbf{P}$ given by

   $$\frac{d\mathbf{P}^{(\lambda)}}{d\mathbf{P}} = e^{\lambda g_1 - \lambda^2/2},$$

   the random Gaussian variables $g_1 - \lambda$ and g_2 are independent standard variables.

   (b) Deduce that for all real-valued y_1, y_2, λ_1 and λ_2, we have

   $$\mathbf{E} \left(\exp(y_1 + \lambda_1 g_1) - \exp(y_2 + \lambda_2 g_2) \right)_+$$
   $$= e^{y_1 + \lambda_1^2/2} N \left(\frac{y_1 - y_2 + \lambda_1^2}{\sqrt{\lambda_1^2 + \lambda_2^2}} \right) - e^{y_2 + \lambda_2^2/2} N \left(\frac{y_1 - y_2 - \lambda_2^2}{\sqrt{\lambda_1^2 + \lambda_2^2}} \right),$$

   where N is the standard normal distribution function.

2. Deduce from the previous question an expression for F using the function N.

3. We set $\tilde{C}_t = e^{-rt} F(t, S_t^1, S_t^2)$. Noticing that

   $$\tilde{C}_t = F(t, \tilde{S}_t^1, \tilde{S}_t^2) = \tilde{\mathbf{E}} \left(e^{-rT} \left(S_T^1 - S_T^2 \right)_+ | \mathcal{F}_t \right),$$

   prove the equality

   $$d\tilde{C}_t = \frac{\partial F}{\partial x_1}(t, \tilde{S}_t^1, \tilde{S}_t^2) \sigma_1 e^{-rt} S_t^1 dW_t^1 + \frac{\partial F}{\partial x_2}(t, \tilde{S}_t^1, \tilde{S}_t^2) \sigma_2 e^{-rt} S_t^2 dW_t^2.$$

   Hint: use the fact that if (X_t) is an Itô process which can be written as $X_t = X_0 + \int_0^t J_s^1 dW_s^1 + \int_0^t J_s^2 dW_s^2 + \int_0^t K_s ds$ and if it is a martingale under $\tilde{\mathbf{P}}$, then $K_t = 0$, $dt d\tilde{\mathbf{P}}$-almost everywhere.

4. Build a hedging scheme for the option.

Problem 4 A study of strategies with consumption

We consider a financial market in which there is one riskless asset, with price $S_t^0 = e^{rt}$ at time t (with $r \geq 0$) and one risky asset, with price S_t at time t. The model is studied on the time interval $[0, T]$ ($0 \leq T < \infty$). In the following, $(S_t)_{0 \leq t \leq T}$ is a stochastic process defined on a probability space $(\Omega, \mathcal{F}, \mathbf{P})$, equipped with a filtration $(\mathcal{F}_t)_{0 \leq t \leq T}$. We assume that $(\mathcal{F}_t)_{0 \leq t \leq T}$ is the natural filtration of a standard Brownian motion $(B_t)_{0 \leq t \leq T}$ and that the process $(S_t)_{0 \leq t \leq T}$ is adapted to this filtration.

I

We want to study strategies in which consumption is allowed. The dynamic of $(S_t)_{0 \leq t \leq T}$ is given by the Black-Scholes model

$$dS_t = S_t(\mu dt + \sigma dB_t),$$

with $\mu \in \mathbb{R}$ and $\sigma > 0$. We note $\mathbf{P}^*$ the probability with density

$$\exp\left(-\theta B_T - \theta^2 T/2\right)$$

with respect to P, with $\theta = (\mu - r)/\sigma$. Under $\mathbf{P}^*$, the process $(W_t)_{0 \leq t \leq T}$, defined by $W_t = (\mu - r)t/\sigma + B_t$, is a standard Brownian motion.

A strategy with consumption is defined by three stochastic processes: $(H_t^0)_{0 \leq t \leq T}$, $(H_t)_{0 \leq t \leq T}$ and $(c(t))_{0 \leq t \leq T}$. H_t^0 and H_t represent respectively the quantities of riskless asset and risky asset held at time t, and $c(t)$ represents the consumption rate at time t. We say that such a strategy is *admissible* if the following conditions hold:

(i) The processes $(H_t^0)_{0 \leq t \leq T}$, $(H_t)_{0 \leq t \leq T}$ and $(c(t))_{0 \leq t \leq T}$ are adapted and satisfy

$$\int_0^T \left(|H_t^0| + H_t^2 + |c(t)|\right) dt < \infty, \quad \text{a.s.}$$

(ii) For all $t \in [0, T]$,

$$H_t^0 S_t^0 + H_t S_t = H_0^0 S_0^0 + H_0 S_0 + \int_0^t H_u^0 dS_u^0 + \int_0^t H_u dS_u - \int_0^t c(u)du, \quad \text{a.s.}$$

(iii) For all $t \in [0, T]$, $c(t) \geq 0$ a.s.

(iv) For all $t \in [0, T]$, the random variable $H_t^0 S_t^0 + H_t S_t$ is non-negative and

$$\sup_{t \in [0,T]} \left(H_t^0 S_t^0 + H_t S_t + \int_0^t c(s)ds\right)$$

is square-integrable under the probability $\mathbf{P}^*$.

1. Let $(H_t^0)_{0 \leq t \leq T}$, $(H_t)_{0 \leq t \leq T}$ and $(c(t))_{0 \leq t \leq T}$ be three adapted processes satisfying condition (i) above. We set $V_t = H_t^0 S_t^0 + H_t S_t$ and $\tilde{V}_t = e^{-rt} V_t$. Then

show that condition (ii) is satisfied if and only if we have, for all $t \in [0, T]$,

$$\tilde{V}_t = V_0 + \int_0^t H_u d\tilde{S}_u - \int_0^t \tilde{c}(u) du, \quad \text{a.s.}$$

with $\tilde{S}_u = e^{-ru} S_u$ and $\tilde{c}(u) = e^{-ru} c(u)$.

2. We suppose that conditions (i) to (iv) are satisfied and we still note $\tilde{V}_t = e^{-rt} V_t = e^{-rt} \left(H_t^0 S_t^0 + H_t S_t \right)$. Prove that the process $(\tilde{V}_t)_{0 \leq t \leq T}$ is a super-martingale under probability $\mathbf{P}^*$.

3. Let $(c(t))_{0 \leq t \leq T}$ be an adapted process with non-negative values such that $\mathbf{E}^* \left(\int_0^T c(t) dt \right)^2 < \infty$ and let $x > 0$. We say that $(c(t))_{0 \leq t \leq T}$ is a budget-feasible consumption process from the initial endowment x if there exist some processes $(H_t^0)_{0 \leq t \leq T}$ and $(H_t)_{0 \leq t \leq T}$ such that conditions (i) to (iv) are satisfied, and furthermore $V_0 = H_0^0 S_0^0 + H_0 S_0 = x$.

   (a) Show that if the process $(c(t))_{0 \leq t \leq T}$ is budget-feasible from the initial endowment x then $\mathbf{E}^* \left(\int_0^T e^{-rt} c(t) dt \right) \leq x$.

   (b) Let $(c(t))_{0 \leq t \leq T}$ be an adapted process, with non-negative values and such that

   $$\mathbf{E}^* \left(\int_0^T c(t) dt \right)^2 < \infty \quad \text{and} \quad \mathbf{E}^* \left(\int_0^T e^{-rt} c(t) dt \right) \leq x.$$

   Prove that $(c(t))_{0 \leq t \leq T}$ is a budget-feasible consumption process with an initial endowment x. Hint: introduce the martingale $(M_t)_{0 \leq t \leq T}$ defined by $M_t = \mathbf{E}^* \left(x + \int_0^T e^{-rs} c(s) ds | \mathcal{F}_t \right)$ and apply the theorem of martingales representation.

   (c) An investor with initial endowment x wants to consume a wealth corresponding to the sale of ρ risky assets by unit of time whenever S_t crosses some barrier K upward (that corresponds to $c(t) = \rho S_t \mathbf{1}_{\{S_t > K\}}$). What conditions on ρ and x are necessary for this consumption process to be budget-feasible?

II

We now suppose that the volatility is stochastic, i.e. that the process $(S_t)_{0 \leq t \leq T}$ is the solution of a stochastic differential equation of the following form:

$$dS_t = S_t (\mu dt + \sigma(t) dB_t), \tag{4.20}$$

where $\mu \in \mathbb{R}$ and $(\sigma(t))_{0 \leq t \leq T}$ is an adapted process, satisfying

$$\forall t \in [0, T] \quad \sigma_1 \leq \sigma(t) \leq \sigma_2,$$

for some constants σ_1 and σ_2 such that $0 < \sigma_1 < \sigma_2$. We consider a European call with maturity T and strike price K on one unit of the risky asset. We know (see Chapter 5) that if the process $(\sigma(t))_{0 \le t \le T}$ is constant (with $\sigma(t) = \sigma$ for any t) the price of the call at time t is $C(t, S_t)$, where the function $C(t, x)$ satisfies

$$
\begin{cases}
\dfrac{\partial C}{\partial t}(t, x) + \dfrac{\sigma^2 x^2}{2} \dfrac{\partial^2 C}{\partial x^2}(t, x) + rx \dfrac{\partial C}{\partial x}(t, x) - rC(t, x) = 0 \\[2ex]
\text{on } [0, T[\times]0, \infty[\\[2ex]
C(T, x) = (x - K)_+.
\end{cases}
$$

We note C_1 the function C corresponding to the case $\sigma = \sigma_1$ and C_2 the function C corresponding to the case $\sigma = \sigma_2$. We want to show that the price of the call at time 0 in the model with stochastic volatility belongs to $[C_1(0, S_0), C_2(0, S_0)]$.

Recall that if $(\theta_t)_{0 \le t \le T}$ is a bounded adapted process, the process $(L_t)_{0 \le t \le T}$ defined by $L_t = \exp\left(\int_0^t \theta_s dB_s - \frac{1}{2} \int_0^t \theta_s^2 ds\right)$ is a martingale.

1. Prove (using the price formulae written as expectations) that the functions $x \mapsto C_1(t, x)$ and $x \mapsto C_2(t, x)$ are convex.

2. Show that the solution of equation (4.20) is given by

$$
S_t = S_0 \exp\left(\mu t + \int_0^t \sigma(s) dB_s - \frac{1}{2} \int_0^t \sigma^2(s) ds\right).
$$

3. Determine a probability $\mathbf{P}^*$ equivalent to $\mathbf{P}$ under which the process defined by $W_t = B_t + \int_0^t (\mu - r)/\sigma(s) ds$ is a standard Brownian motion.

4. Explain why the price of the call at time 0 is given by

$$
C_0 = \mathbf{E}^* \left(e^{-rT}(S_T - K)_+\right).
$$

5. We set $\tilde{S}_t = e^{-rt} S_t$. Show that $\mathbf{E}^* \left(\tilde{S}_t^2\right) \le S_0^2 e^{\sigma_2^2 t}$.

6. Prove that the process defined by

$$
M_t = \int_0^t e^{-ru} \frac{\partial C_1}{\partial x}(u, S_u) \sigma(u) S_u dW_u
$$

is a martingale under probability $\mathbf{P}^*$.

7. Using Itô formula and Questions 1 and 6, show that $e^{-rt} C_1(t, S_t)$ is a submartingale under probability $\mathbf{P}^*$. Deduce that $C_1(0, S_0) \le C_0$.

8. Derive the inequality $C_0 \le C_2(0, S_0)$.

Problem 5 Compound option

We consider a financial market offering two investment opportunities. The first traded security is a riskless asset whose price is equal to $S_t^0 = e^{rt}$ at time t (with $r \ge 0$) and the second security is risky and its price is denoted by S_t at time $t \in [0, T]$. Let $(S_t)_{0 \le t \le T}$ be a stochastic process defined on a probability space $(\Omega, \mathcal{F}, \mathbf{P})$, equipped with a filtration $(\mathcal{F}_t)_{0 \le t \le T}$. We assume that $(\mathcal{F}_t)_{0 \le t \le T}$ is the natural filtration generated by a standard Brownian motion $(B_t)_{0 \le t \le T}$ and that $(S_t)_{0 \le t \le T}$ follows a Black-Scholes model

$$dS_t = S_t(\mu dt + \sigma dB_t),$$

with $\mu \in \mathbb{R}$ and $\sigma > 0$.

We want to study an example of compound option. We consider a call option with maturity $T_1 \in]0, T[$ and strike price K_1 on a call of maturity T and strike price K. The value of this option at time T_1 is equal to

$$h = (C(T_1, S_{T_1}) - K_1)_+,$$

where $C(t, x)$ is the price of the underlying call, given by the Black-Scholes formula.

1.

(a) Graph function $x \mapsto C(T_1, x)$. Show that the line $y = x - Ke^{-r(T-T_1)}$ is an asymptote (hint: use the put/call parity).

(b) Show that the equation $C(T_1, x) = K_1$ has a unique solution x_1.

2. Show that at time $t < T_1$, the compound option is worth $G(T_1 - t, S_t)$, where G is defined by

$$G(\theta, x) = \mathbf{E}\left[e^{-r\theta}\left(C\left(T_1, xe^{\left(r - \frac{\sigma^2}{2}\right)\theta + \sigma\sqrt{\theta}g}\right) - K_1\right)^+\right],$$

with g being a standard normal random variable.

3.

(a) Show that $x \mapsto G(\theta, x)$ is an increasing convex function.

(b) We now want to compute G explicitly. Let us denote by N the standard cumulative normal distribution. Prove that

$$G(\theta, x) = \mathbf{E}\left[e^{-r\theta}C\left(T_1, xe^{(r - \sigma^2/2)\theta + \sigma\sqrt{\theta}g}\right)1_{\{g > -d\}}\right] - K_1 e^{-r\theta}N(d),$$

where

$$d = \frac{\log(x/x_1) + (r - \sigma^2/2)\,\theta}{\sigma\sqrt{\theta}}.$$

(c) Show that if g_1 is a standard normal variable independent of g, we can write $\theta_1 = T - T_1$ and characterise G by,

$$G(\theta, x) + K_1 e^{-r\theta}N(d)$$
$$= \mathbf{E}\left[\left(xe^{\sigma(\sqrt{\theta}g + \sqrt{\theta_1}g_1) - \frac{\sigma^2}{2}(\theta + \theta_1)} - Ke^{-r(\theta + \theta_1)}\right)1_A\right],$$

where the event A is defined by

$$A = \left\{\sigma(\sqrt{\theta}g + \sqrt{\theta_1}g_1) > - \left(\log(x/K_1) + \left(r - \frac{\sigma^2}{2}\right)(\theta + \theta_1)\right)\right.$$
$$\left. \text{and } g > -d\right\}.$$

(d) From this, derive a formula for $G(\theta, x)$ in terms of N and N_2 the two-dimensional cumulative normal distribution defined by

$$N_2(y, y_1, \rho) = \mathbf{P}(g < y, g + \rho g_1 < y_1) \quad \text{for} \quad y, y_1, \rho \in \mathbb{R}.$$

4. Show that we can replicate the compound option payoff by trading the underlying call and the riskless bond.

Problem 6 Behaviour of the critical price close to maturity

We consider an American put maturing at T with strike price K on a share of risky asset S. In the Black-Scholes model, its value at time $t < T$ is equal to $P(t, S_t)$, when P is defined by

$$P(t, x) = \sup_{\tau \in \mathcal{T}_{0, T-t}} \mathbf{E}^* \left(K e^{-r\tau} - x e^{\sigma W_\tau - \frac{\sigma^2}{2}\tau} \right)^+,$$

$\mathcal{T}_{0, T-t}$ is the set of stopping times with values in $[0, T - t]$ and $(W_t)_{0 \le t \le T}$ is a standard $\mathbf{P}^*$-Brownian motion. We also assume that $r > 0$. For $t \in [0, T[$, we denote by $s(t)$ the critical price defined as

$$s(t) = \inf\{x > 0 \mid P(t, x) > K - x\}.$$

we recall that $\lim_{t \to T} s(t) = K$.

1. Let P_e be the function pricing the European put with maturity T and strike price K

$$P_e(t, x) = \mathbf{E} \left(e^{-r(T-t)} K - x e^{\sigma \sqrt{T-t} g - \frac{\sigma^2}{2}(T-t)} \right)^+,$$

   where g is a standard normal variable. Show that if $t \in [0, T[$, the equation $P_e(t, x) = K - x$ has a unique solution in $]0, K[$. Let us call it $s_e(t)$.

2. Show that $s(t) \le s_e(t)$, for any $t \in [0, T[$.

3. Show that

$$\liminf_{t \to T} \frac{K - s_e(t)}{\sqrt{T - t}} \ge \mathbf{E} \left(\liminf_{t \to T} \frac{K - s_e(t)}{\sqrt{T - t}} - \sigma K g \right)^+.$$

   We shall need Fatou lemma: for any sequence $(X_n)_{n \in \mathbb{N}}$ of non-negative random variables, $\mathbf{E}(\liminf_{n \to \infty} X_n) \le \liminf_{n \to \infty} \mathbf{E}(X_n)$.

4.

   (a) Show that for any real number η,

   $$\mathbf{E}(\eta - K\sigma g)^+ > \eta.$$

   (b) Deduce that

   $$\lim_{t \to T} \frac{K - s_e(t)}{\sqrt{T - t}} = \lim_{t \to T} \frac{K - s(t)}{\sqrt{T - t}} = +\infty.$$

Problem 7 Asian option

We consider a financial market offering two investment opportunities. The first traded security is a riskless asset whose price is equal to $S_t^0 = e^{rt}$ at time t (with $r \geq 0$) and the second security is risky and its price is denoted by S_t at time $t \in [0, T]$. Let $(S_t)_{0 \leq t \leq T}$ be a stochastic process defined on a probability space $(\Omega, \mathcal{F}, \mathbf{P})$, equipped with a filtration $(\mathcal{F}_t)_{0 \leq t \leq T}$. We assume that $(\mathcal{F}_t)_{0 \leq t \leq T}$ is the natural filtration generated by a standard Brownian motion $(B_t)_{0 \leq t \leq T}$ and that $(S_t)_{0 \leq t \leq T}$ follows a Black-Scholes model

$$dS_t = S_t(\mu dt + \sigma dB_t),$$

with $\mu \in \mathbb{R}$ and $\sigma > 0$. We shall denote by $\mathbf{P}^*$ the probability measure with density $\exp\left(-\theta B_T - \theta^2 T/2\right)$ with respect to $\mathbf{P}$, where $\theta = (\mu - r)/\sigma$. Under $\mathbf{P}^*$, the process $(W_t)_{0 \leq t \leq T}$, defined by $W_t = (\mu - r)t/\sigma + B_t$ is a standard Brownian motion.

We are going to study the option whose payoff is equal to

$$h = \left(\frac{1}{T} \int_0^T S_t dt - K\right)^+,$$

where K is a positive constant.

I

1. Explain briefly why the Asian option price at time t ($t \leq T$) is given by

$$V_t = \mathbf{E}^* \left[e^{-r(T-t)} \left(\frac{1}{T} \int_0^T S_u du - K\right)^+ \middle| \mathcal{F}_t \right].$$

2. Show that on the event $\left\{ \int_0^t S_u du \geq KT \right\}$, we have

$$V_t = \frac{e^{-r(T-t)}}{T} \int_0^t S_u du + \frac{1 - e^{-r(T-t)}}{rT} S_t - K e^{-r(T-t)}.$$

3. We define $\tilde{S}_t = e^{-rt} S_t$, for $t \in [0, T]$.

   (a) Derive the inequality

   $$\mathbf{E}^* \left(\tilde{S}_t - K e^{-rT}\right)^+ \leq \mathbf{E}^* \left[e^{-rT} (S_T - K)^+\right].$$

   (Use conditional expectations given $\mathcal{F}_t$).

   (b) Deduce that

   $$V_0 \leq \mathbf{E}^* \left[e^{-rT} (S_T - K)^+\right],$$

   i.e. the Asian option price is smaller than its European counterpart.

   (c) For $t \leq u$, we denote by $C_{t,u}$ the value at time t of a European call maturing at time u with strike price K. Prove the following inequality

$$V_t \leq \frac{e^{-r(T-t)}t}{T} \left(\frac{1}{t} \int_0^t S_u du - K \right)^+ + \frac{1}{T} \int_t^T e^{-r(T-u)} C_{t,u} du.$$

II

We denote by $(\xi_t)_{0 \leq t \leq T}$ the process defined by

$$\xi_t = \frac{1}{S_t} \left(\frac{1}{T} \int_0^t S_u du - K \right).$$

1. Show that $(\xi_t)_{0 \leq t \leq T}$ is the solution of the following stochastic differential equation:

$$d\xi_t = \left(\frac{1}{T} + (\sigma^2 - r)\xi_t \right) dt - \sigma \xi_t dW_t.$$

2.

(a) Show that

$$V_t = e^{-r(T-t)} S_t \mathbf{E}^* \left[\left(\xi_t + \frac{1}{T} \int_t^T S_u^t du \right)^+ |\mathcal{F}_t \right],$$

with $S_u^t = \exp \left((r - \sigma^2/2)(u-t) + \sigma(W_u - W_t) \right).$

(b) Conclude that $V_t = e^{-r(T-t)} S_t F(t, \xi_t)$, with

$$F(t, \xi) = \mathbf{E}^* \left(\xi + \frac{1}{T} \int_t^T S_u^t du \right)^+.$$

3. Find a replicating strategy to hedge the Asian option. We shall assume that the function F introduced earlier is of class C^2 on $[0, T[\times \mathbb{R}$ and we shall use Itô formula.

III

The purpose of this section is to suggest an approximation of V_0 obtained by considering the geometric average as opposed to the arithmetic one. We define

$$\hat{V}_0 = e^{-rT} \mathbf{E}^* \left(\exp \left(\frac{1}{T} \int_0^T \ln(S_t) dt \right) - K \right)^+.$$

1. Show that $V_0 \geq \hat{V}_0$.

2.

(a) Show that under measure $\mathbf{P}^*$, the random variable $\int_0^T W_t dt$ is normal with zero mean and a variance equal to $T^3/3$.

(b) Deduce that

$$\hat{V}_0 = e^{-rT} \mathbf{E} \left(S_0 \exp \left((r - \sigma^2/2)(T/2) + \sigma\sqrt{T/3}g \right) - K \right)^+,$$

where g is a standard normal variable. Give a closed-form formula for $\hat{V}_0$ in terms of the normal distribution function.

3. Prove the inequality

$$V_0 - \hat{V}_0 \leq S_0 e^{-rT} \left(\frac{e^{rT} - 1}{rT} - \exp\left((rT/2) - \sigma^2 T/12 \right) \right).$$

5

Option pricing and partial differential equations

In the previous chapter, we saw how we could derive a closed-form formula for the price of a European option in the Black-Scholes environment. But, if we are working with more complex models or even if we want to price American options, we are not able to find such explicit expressions. That is why we will often require numerical methods. The purpose of this chapter is to give an introduction to some concepts useful for computations.

Firstly, we shall show how the problem of European option pricing is related to a parabolic partial differential equation (PDE). This link is based on the concept of the infinitesimal generator of a diffusion. We shall also address the problem of solving the PDE numerically.

The pricing of American options is rather difficult and we will not attempt to address it in its whole generality. We shall concentrate on the Black-Scholes model and, in particular, we shall underline the natural duality between the Snell envelope and a parabolic system of partial differential *inequalities*. We shall also explain how we can solve this kind of system numerically.

We shall only use classical numerical methods and therefore we will just recall the few results that we need. However, an introduction to numerical methods to solve parabolic PDEs can be found in Ciarlet and Lions (1990) or Raiviart and Thomas (1983).

5.1 European option pricing and diffusions

In a Black-Scholes environment, the European option price is given by

$$V_t = \mathbf{E}\left(e^{-r(T-t)}f(S_T)|\mathcal{F}_t\right)$$

with $f(x) = (x - K)_+$ (for a call), $(K - x)_+$ (for a put) and

$$S_T = x_0 e^{(r-\sigma^2/2)T + \sigma W_T}.$$

In fact, we should point out that the pricing of a European option is only a particular case of the following problem. Let $(X_t)_{t\geq 0}$ be a diffusion in $\mathbb{R}$, solution of

$$dX_t = b(t, X_t)\, dt + \sigma(t, X_t)\, dW_t \tag{5.1}$$

where b and σ are real-valued functions satisfying the assumptions of Theorem 3.5.3 in Chapter 3 and $r(t, x)$ is a bounded continuous function modelling the riskless interest rate. We generally want to compute

$$V_t = \mathbf{E}\left(e^{-\int_t^T r(s, X_s)ds} f(X_T)|\mathcal{F}_t\right).$$

In the same way as in the Black-Scholes model, V_t can be written as

$$V_t = F(t, X_t)$$

where

$$F(t, x) = \mathbf{E}\left(e^{-\int_t^T r(s, X_s^{t,x})ds} f(X_T^{t,x})\right),$$

and $X_s^{t,x}$ is the solution of (5.1) starting from x at time t. Intuitively

$$F(t, x) = \mathbf{E}\left(e^{-\int_t^T r(s, X_s)ds} f(X_T)|X_t = x\right).$$

Mathematically, this result is a consequence of Theorem 3.5.9 in Chapter 3. The computation of V_t is therefore equivalent to the computation of $F(t, x)$. Under some regularity assumptions that we shall specify, this function $F(t, x)$ is the unique solution of the following partial differential equation

$$\begin{cases} \forall x \in \mathbb{R} \quad u(T, x) = f(x) \\ \\ (\partial u/\partial t + A_t u - ru)(t, x) = 0 \quad \forall (t, x) \in [0, T] \times \mathbb{R} \end{cases} \tag{5.2}$$

where

$$(A_t f)(x) = \frac{\sigma^2(t, x)}{2} f''(x) + b(t, x)f'(x).$$

Before we prove this result, let us explain why the operator A_t appears naturally when we solve *stochastic differential equations*.

5.1.1 Infinitesimal generator of a diffusion

We assume that b and σ are time independent and we denote by $(X_t)_{t\geq 0}$ the solution of

$$dX_t = b(X_t)\, dt + \sigma(X_t)\, dW_t. \tag{5.3}$$

Proposition 5.1.1 *Let f be a C^2 function with bounded derivatives and A be the differential operator that maps a C^2 function f to Af such that*

$$(Af)(x) = \frac{\sigma^2(x)}{2} f''(x) + b(x)f'(x).$$

Then, the process $M_t = f(X_t) - \int_0^t Af(X_s)ds$ is an F_t-martingale.

Proof. Itô formula yields

$$f(X_t) = f(X_0) + \int_0^t f'(X_s)dX_s + \frac{1}{2}\int_0^t f''(X_s)\sigma^2(X_s)ds.$$

Hence

$$
\begin{aligned}
f(X_t) &= f(X_0) + \int_0^t f'(X_s)\sigma(X_s)dW_s \\
&+ \int_0^t \left[\frac{1}{2}\sigma^2(X_s)f''(X_s) + b(X_s)f'(X_s)\right] ds
\end{aligned}
$$

and the result follows from the fact that the stochastic integral $\int_0^t f'(X_s)\sigma(X_s)dW_s$ is a martingale. Indeed, according to Theorem 3.5.3 and since $|\sigma(x)|$ is dominated by $K(1+|x|)$, we obtain

$$
\mathbf{E} \left(\int_0^t |f'(X_s)|^2|\sigma(X_s)|^2 ds \right)
$$

$$
\leq KT \sup_{x\in\mathbb{R}} |f'(x)|^2 \left(1 + \mathbf{E}\left(\sup_{s\leq T} |X_s|^2 \right) \right) < +\infty.
$$

$\square$

Remark 5.1.2 If we denote by X_t^x the solution of (5.3) such that $X_0^x = x$, it follows from Proposition 5.1.1 that

$$
\mathbf{E}\left(f\left(X_t^x\right)\right) = f(x) + \mathbf{E}\left(\int_0^t Af\left(X_s^x\right) ds \right).
$$

Moreover, since the derivatives of f are bounded by a constant K_f and since $|b(x)| + |\sigma(x)| \leq K(1+|x|)$ we can say that

$$
\mathbf{E}\left(\sup_{s\leq T} |Af(X_s^x)| \right) \leq K_f' \left(1 + \mathbf{E}\left(\sup_{s\leq T} |X_s^x|^2 \right) \right) < +\infty.
$$

Therefore, since $x \mapsto Af(x)$ and $s \mapsto X_s^x$ are continuous, the Lebesgue theorem is applicable and yields

$$
\frac{d}{dt} \mathbf{E}\left(f\left(X_t^x\right)\right)\Big|_{t=0} = \lim_{t\to 0} \mathbf{E}\left(\frac{1}{t}\int_0^t Af(X_s^x)ds \right) = Af(x).
$$

The differential operator A is called the *infinitesimal generator* of the diffusion (X_t). The reader can refer to Bouleau (1988) or Revuz and Yor (1990) to study some properties of the infinitesimal generator of a diffusion.

The Proposition 5.1.1 can also be extended to the time-dependent case. We assume that b and σ satisfy the assumptions of Theorem 3.5.3 in Chapter 3 which guarantee the existence and uniqueness of a solution of equation (5.1).

Proposition 5.1.3 *If $u(t,x)$ is a $C^{1,2}$ function with bounded derivatives in x and if X_t is a solution of (5.1), the process*

$$M_t = u(t, X_t) - \int_0^t \left(\frac{\partial u}{\partial t} + A_s u \right) (s, X_s) ds$$

is a martingale. Here, A_s is the operator defined by

$$(A_s u)(x) = \frac{\sigma^2(s,x)}{2} \frac{\partial^2 u}{\partial x^2} + b(s,x) \frac{\partial u}{\partial x}.$$

The proof is very similar to that of Proposition 5.1.1: the only difference is that we apply the Itô formula for a function of time and an Itô process (see Theorem 3.4.10).

In order to deal with discounted quantities, we state a slightly more general result in the following proposition.

Proposition 5.1.4 *Under the assumptions of Proposition 5.1.3, and if $r(t,x)$ is a bounded continuous function defined on $\mathbb{R}^+ \times \mathbb{R}$, the process*

$$M_t = e^{-\int_0^t r(s,X_s)ds} u(t, X_t) - \int_0^t e^{-\int_0^s r(v,X_v)dv} \left(\frac{\partial u}{\partial t} + A_s u - ru \right) (s, X_s) ds$$

is a martingale.

Proof. This proposition can be proved by using the integration by parts formula to differentiate the product (see Proposition 3.4.12 in Chapter 3)

$$e^{-\int_0^t r(s,X_s)ds} u(t, X_t),$$

and then applying Itô formula to the process $u(t, X_t)$.   □

This result is still true in a multidimensional model. Let us consider the stochastic differential equation

$$\begin{cases} dX_t^1 &= b^1(t, X_t) dt + \sum_{j=1}^p \sigma_{1j}(t, X_t) dW_t^j \\ \vdots & \qquad \vdots \\ dX_t^n &= b^n(t, X_t) dt + \sum_{j=1}^p \sigma_{nj}(t, X_t) dW_t^j. \end{cases} \qquad (5.4)$$

We assume that the assumptions of Theorem 3.5.5 are still satisfied. For any time t we define the following differential operator A_t which maps a C^2 function from $\mathbb{R}^n$ to $\mathbb{R}$ to a function characterised by

$$(A_t f)(x) = \frac{1}{2} \sum_{i,j=1}^n a_{i,j}(t,x) \frac{\partial^2 f}{\partial x_i \partial x_j}(x) + \sum_{j=1}^n b_j(t,x) \frac{\partial f}{\partial x_j}(x),$$

where $(a_{ij}(t,x))$ is the matrix of components

$$a_{ij}(t,x) = \sum_{k=1}^p \sigma_{ik}(t,x) \sigma_{jk}(t,x)$$

In other words $a(t,x) = \sigma(t,x)\sigma^*(t,x)$ where σ^* is the transpose of $\sigma(t,x) = (\sigma_{ij}(t,x))$.

Proposition 5.1.5 *If* (X_t) *is a solution of system (5.4) and* $u(t,x)$ *is a real-valued function of class* $C^{1,2}$ *defined on* $\mathbb{R}^+ \times \mathbb{R}^n$ *with bounded derivatives in* x *and, also,* $r(t,x)$ *is a continuous bounded function defined on* $\mathbb{R}^+ \times \mathbb{R}$, *then the process*

$$M_t = e^{-\int_0^t r(s,X_s)ds} u(t,X_t) - \int_0^t e^{-\int_0^s r(v,X_v)dv} \left(\frac{\partial u}{\partial t} + A_s u - ru \right)(s,X_s)ds$$

is a martingale.

The proof is based on the multidimensional Itô formula stated page 48.

Remark 5.1.6 The differential operator $\partial/\partial t + A_t$ is sometimes called the *Dynkin operator of the diffusion.*

5.1.2 Conditional expectations and partial differential equations

In this section, we want to emphasise the link between pricing a European option and solving a parabolic partial differential equation. Let us consider $(X_t)_{t\geq 0}$ a solution of system (5.4), $f(x)$ a function from $\mathbb{R}^n$ to $\mathbb{R}$, and $r(t,x)$ a bounded continuous function. We want to compute

$$V_t = \mathbf{E}\left(e^{-\int_t^T r(s,X_s)ds} f(X_T) | \mathcal{F}_t \right).$$

In a similar way, as in the scalar case, we can prove that

$$V_t = F(t,X_t),$$

where

$$F(t,x) = \mathbf{E}\left(e^{-\int_t^T r(s,X_s^{t,x})ds} f(X_T^{t,x}) \right),$$

when we denote by $X^{t,x}$ the unique solution of (5.4) starting from x at time t.

The following result characterises the function F as a solution of a partial differential equation.

Theorem 5.1.7 *Let* u *be a* $C^{1,2}$ *function with a bounded derivative in* x *defined on* $[0,T] \times \mathbb{R}^n$. *If* u *satisfies*

$$\forall x \in \mathbb{R}^n \quad u(T,x) = f(x),$$

and

$$\left(\frac{\partial u}{\partial t} + A_t u - ru \right)(t,x) = 0 \quad \forall(t,x) \in [0,T] \times \mathbb{R}^n,$$

then

$$\forall(t,x) \in [0,T] \times \mathbb{R}^n \quad u(t,x) = F(t,x) = \mathbf{E}\left(e^{-\int_t^T r(s,X_s^{t,x})ds} f(X_T^{t,x}) \right).$$

Proof. Let us prove the equality $u(t, x) = F(t, x)$ at time $t = 0$. By Proposition 5.1.5, we know that the process

$$M_t = e^{-\int_0^t r(s, X_s^{0,x}) ds} u(t, X_t^{0,x})$$

is a martingale. Therefore the relation $\mathbf{E}(M_0) = \mathbf{E}(M_T)$ yields

$$
\begin{aligned}
u(0, x) &= \mathbf{E}\left(e^{-\int_0^T r(s, X_s^{0,x}) ds} u(T, X_T^{0,x})\right) \\
&= \mathbf{E}\left(e^{-\int_0^T r(s, X_s^{0,x}) ds} f(X_T^{0,x})\right)
\end{aligned}
$$

since $u(T, x) = f(x)$. The proof runs similarly for $t > 0$. $\qquad\square$

Remark 5.1.8 Obviously, Theorem 5.1.7 suggests the following method to price the option. In order to compute

$$F(t, x) = \mathbf{E}\left(e^{-\int_t^T r(s, X_s^{t,x}) ds} f(X_T^{t,x})\right)$$

for a given f, we just need to find u such that

$$
\begin{cases}
\dfrac{\partial u}{\partial t} + A_t u - ru = 0 & \text{in } [0, T] \times \mathbb{R}^n \\[2mm]
u(T, x) = f(x), \ \forall x \in \mathbb{R}^n.
\end{cases}
\tag{5.5}
$$

Problem (5.5) is a parabolic equation with a *final* condition (as soon as the function $u(T, .)$ is given).

For the problem to be well defined, we need to work in a very specific function space (see Raviart and Thomas (1983)). Then we can apply some theorems of existence and uniqueness, and if the solution u of (5.5) is smooth enough to satisfy the assumptions of Proposition 5.1.4 we can conclude that $F = u$. Generally speaking, we shall impose some regularity assumptions on the parameters b and σ and the operator A_t will need to be elliptic, i.e.

$$\exists C > 0, \ \forall (t, x) \in [0, T] \times \mathbb{R}^n$$

$$\forall (\xi_1, \ldots, \xi_n) \in \mathbb{R}^n \quad \sum_{ij} a_{ij}(t, x) \xi_i \xi_j \geq C \left(\sum_{i=1}^n \xi_i^2\right). \tag{5.6}$$

5.1.3 Application to the Black-Scholes model

We are working under probability $\mathbf{P}^*$. The process $(W_t)_{t \geq 0}$ is a standard Brownian motion and the asset price S_t satisfies

$$dS_t = S_t \left(r dt + \sigma dW_t\right).$$

The operator A_t is now time independent and is equal to

$$A_t = A^{bs} = \frac{\sigma^2}{2}x^2\frac{\partial^2}{\partial x^2} + rx\frac{\partial}{\partial x}.$$

It is straightforward to check that the call price given by $F(t,x) = xN(d_1) - Ke^{-r(T-t)}N(d_1 - \sigma\sqrt{T-t})$ with

$$d_1 = \frac{\log(x/K) + (r + \sigma^2/2)(T-t)}{\sigma\sqrt{T-t}}$$

$$N(d) = \frac{1}{\sqrt{2\pi}}\int_{-\infty}^{d} e^{-x^2/2}dx,$$

is solution of the equation

$$\begin{cases} \dfrac{\partial u}{\partial t} + A^{bs}u - ru = 0 & \text{in } [0,T]\times]0,+\infty[\\ \\ u(T,x) = (x-K)_+, \ \forall x \in]0,+\infty[. \end{cases}$$

The same type of result holds for the put.

Note that the operator A^{bs} does not satisfy the ellipticity condition (5.6). However, the trick is to consider the diffusion $X_t = \log(S_t)$, which is solution of

$$dX_t = \left(r - \frac{\sigma^2}{2}\right)dt + \sigma dW_t,$$

since $S_t = S_0 e^{(r-\sigma^2/2)t+\sigma W_t}$. Its infinitesimal generator can be written as

$$A^{bs-\log} = \frac{\sigma^2}{2}\frac{\partial^2}{\partial x^2} + \left(r - \frac{\sigma^2}{2}\right)\frac{\partial}{\partial x}.$$

It is clearly elliptic because $\sigma^2 > 0$ and, moreover, it has constant coefficients. We write

$$\tilde{A}^{bs-\log} = \frac{\sigma^2}{2}\frac{\partial^2}{\partial x^2} + \left(r - \frac{\sigma^2}{2}\right)\frac{\partial}{\partial x} - r. \tag{5.7}$$

The connection between the parabolic problem associated to $\tilde{A}^{bs-\log}$ and the computation of the price of an option in the Black-Scholes model can be highlighted as follows: if we want to compute the price $F(t,x)$ at time t and for a spot price x of an option paying off $f(S_T)$ at time T, we need to find a regular solution v of

$$\begin{cases} \dfrac{\partial v}{\partial t}(t,x) + \tilde{A}^{bs-\log}v(t,x) = 0 & \text{in } [0,T]\times\mathbb{R} \\ \\ v(T,x) = f(e^x), \ \forall x \in \mathbb{R}, \end{cases} \tag{5.8}$$

then $F(t,x) = v(t,\log(x))$.

5.1.4 Partial differential equations on a bounded open set and computation of expectations

Throughout the rest of this section, we shall assume that there is only one asset and that $b(x)$, $\sigma(x)$ and $r(x)$ are all time independent. $r(x)$ is the riskless rate and A is the differential operator defined by

$$(Af)(x) = \frac{1}{2}\sigma(x)^2\frac{\partial^2 f(x)}{\partial x^2} + b(x)\frac{\partial f(x)}{\partial x}.$$

We denote by $\tilde{A}$ the *discount* operator such that $\tilde{A}f(x) = Af(x) - r(x)f(x)$. Equation (5.5) becomes

$$\begin{cases} \dfrac{\partial u}{\partial t}(t,x) + \tilde{A}u(t,x) = 0 \quad \text{on } [0,T] \times \mathbb{R} \\[2mm] u(T,x) = f(x), \ \forall x \in \mathbb{R}. \end{cases} \tag{5.9}$$

If we want to solve problem (5.9) on $\mathcal{O} =]a,b[$ as opposed to $\mathbb{R}$, we need to consider boundary conditions at a and b. We are going to concentrate on the case when the function takes the value zero on the boundaries. These are the so-called Dirichlet boundary conditions. The problem to be solved is then

$$\begin{cases} \dfrac{\partial u}{\partial t}(t,x) + \tilde{A}u(t,x) = 0 \quad \text{on } [0,T] \times \mathcal{O} \\[2mm] u(t,a) = u(t,b) = 0 \quad \forall t \leq T \\[2mm] u(T,x) = f(x) \quad \forall x \in \mathcal{O}. \end{cases} \tag{5.10}$$

As we are about to explain, a regular solution of (5.10) can also be expressed in terms of the diffusion $X^{t,x}$ which is the solution of (5.3) starting at x at time t.

Theorem 5.1.9 *Let u be a $C^{1,2}$ function with bounded derivative in x that satisfies equation (5.10). We then have*

$$\forall(t,x) \in [0,T] \times \mathcal{O}, \ u(t,x) = \mathbf{E}\left(1_{\{\forall s \in [t,T], \, X_s^{t,x} \in \mathcal{O}\}}e^{-\int_t^T r(X_s^{t,x})ds}f(X_T^{t,x})\right)$$

Proof. We shall prove the result for $t = 0$ since the argument is similar at other times. There exists an extension of the function u from $[0,T] \times \mathcal{O}$ to $[0,T] \times \mathbb{R}$ that is still of class $C^{1,2}$. We shall continue to denote by u such an extension. From Proposition 5.1.4, we know that

$$\begin{aligned} M_t &= e^{-\int_0^t r(X_s^{0,x})ds}u(t,X_t^{0,x}) \\ &\quad - \int_0^t e^{-\int_0^s r(X_v^{0,x})dv}\left(\frac{\partial u}{\partial t} + Au - ru\right)(s,X_s^{0,x})ds \end{aligned}$$

is a martingale. Moreover

$$\tau^x = \inf\left\{0 \leq s \leq T, \ X_s^{0,x} \notin \mathcal{O}\right\} \text{ or } T \text{ if this set is empty}$$

is a bounded stopping time, because $\tau^x = T_a^x \wedge T_b^x \wedge T$ where

$$T_l^x = \inf \left\{ 0 \le s \le T, \, X_s^{t,x} = l \right\}$$

and indeed T_l^x is a stopping time according to Proposition 3.3.6. By applying the optional sampling theorem between 0 and τ^x, we get $\mathbf{E}(M_0) = \mathbf{E}(M_{\tau^x})$, thus by noticing that if $s \in [0, \tau^x]$, $Af(X_s^{0,x}) = 0$, it follows that

$$
\begin{aligned}
u(0,x) &= \mathbf{E} \left(e^{-\int_0^{\tau^x} r(s, X_s^{0,x}) ds} u(\tau^x, X_{\tau^x}^{0,x}) \right) \\
&= \mathbf{E} \left(1_{\{\forall s \in [t,T], \, X_s^{t,x} \in \mathcal{O}\}} e^{-\int_0^T r(s, X_s^{0,x}) ds} u(T, X_T^{0,x}) \right) \\
&\quad + \mathbf{E} \left(1_{\{\exists s \in [t,T], \, X_s^{t,x} \notin \mathcal{O}\}} e^{-\int_0^{\tau^x} r(s, X_s^{0,x}) ds} u(\tau^x, X_{\tau^x}^{0,x}) \right).
\end{aligned}
$$

Furthermore, $f(x) = u(T, x)$ and $u(\tau^x, X_{\tau^x}^{0,x}) = 0$ on the event

$$\left\{ \exists s \in [t, T], \, X_s^{t,x} \notin \mathcal{O} \right\};$$

consequently

$$u(0,x) = \mathbf{E} \left(1_{\{\forall s \in [t,T], \, X_s^{t,x} \in \mathcal{O}\}} e^{-\int_0^T r(s, X_s^{0,x}) ds} f(X_T^{0,x}) \right).$$

That completes the proof for $t = 0$. □

Remark 5.1.10 An option on the $\mathcal{F}_T$-measurable random variable

$$1_{\{\forall s \in [t,T], \, X_s^{t,x} \in \mathcal{O}\}} e^{-\int_t^T r(X_s^{t,x}) ds} f(X_T^{t,x})$$

is called *extinguishable*. Indeed, as soon as the asset price exits the open set $\mathcal{O}$, the option becomes worthless. In the Black-Scholes model, if $\mathcal{O}$ is of the form $]0, l[$ or $]l, +\infty[$ we are able to compute explicit formulae for the option price (see Cox and Rubinstein (1985) and exercise 27 for the pricing of *Down and Out* options).

5.2 Solving parabolic equations numerically

We saw under which conditions the option price coincided with the solution of the partial differential equation (5.9). We now want to address the problem of solving a PDE such as (5.9) numerically and we shall see how we can approximate its solution using the so-called *finite difference* method. This method is obviously useless in the Black-Scholes model since we are able to derive a closed-form solution, but it proves to be useful when we are dealing with more general diffusion models. We shall only state the most important results, but the reader can refer to Glowinsky, Lions and Trémolières (1976) or Raviart and Thomas (1983) for a detailed analysis.

5.2.1 Localisation

Problem (5.9) is set on $\mathbb{R}$. In order to discretise, we will have to work on a bounded open set $\mathcal{O}_l =]-l, l[$, where l is a constant to be chosen carefully in order to optimise the algorithm. We also need to specify the boundary conditions (i.e. at l and $-l$). Typically, we shall impose Dirichlet conditions (i.e. $u(l) = u(-l) = 0$ or some more relevant constants) or Neumann conditions (i.e. $(\partial u / \partial x)(l)$, $(\partial u / \partial x)(-l)$). If we specify Dirichlet boundary conditions, the PDE becomes

$$\begin{cases} \dfrac{\partial u(t,x)}{\partial t} + \tilde{A}u(t,x) = 0 \text{ on } [0,T] \times \mathcal{O}_l \\[2mm] u(t,l) = u(t,-l) = 0 \text{ if } t \in [0,T] \\[2mm] u(T,x) = f(x) \text{ if } x \in \mathcal{O}_l. \end{cases}$$

We are going to show how we can estimate the error that we make if we restrict our state space to $\mathcal{O}_l$. We shall work in a Black-Scholes environment and, thus, the logarithm of the asset price solves the following stochastic differential equation

$$dX_t = (r - \sigma^2/2)dt + \sigma dW_t.$$

We want to compute the price of an option whose payoff can be written as $f(S_T) = f(S_0 e^{X_T})$. We write $\bar{f}(x) = f(e^x)$. To simplify, we adopt Dirichlet boundary conditions. We can prove in that case that the solution u of (5.9) and the solutions u_l of (5.10) are smooth enough to be able to say that

$$u(t,x) = \mathbf{E}\left(e^{-r(T-t)}\bar{f}(X_T^{t,x})\right)$$

and

$$u_l(t,x) = \mathbf{E}\left(1_{\{\forall s \in [t,T],\, |X_s^{t,x}| < l\}} e^{-r(T-t)}\bar{f}(X_T^{t,x})\right)$$

where $X_s^{t,x} = x\exp((r - \sigma^2/2)(s-t) + \sigma(W_s - W_t))$. We assume that the function f (hence $\bar{f}$) is bounded by a constant M and that $r \geq 0$. Then, it is easy to show that

$$|u(t,x) - u_l(t,x)| \leq M\mathbf{P}\left(\exists s \in [t,T], |X_s^{t,x}| \geq l\right).$$

If we call $r' = r - \sigma^2/2$

$$\{\exists s \in [t,T], |X_s^{t,x}| \geq l\} \subset \left\{\sup_{t \leq s \leq T} |x + r'(s-t) + \sigma(W_s - W_t)| \geq l\right\}$$

$$\subset \left\{\sup_{t \leq s \leq T} |x + \sigma(W_s - W_t)| \geq l - |r'T|\right\}.$$

Thus

$$|u(t,x) - u_l(t,x)| \leq MP\left(\sup_{t \leq s \leq T}|x + \sigma(W_s - W_t)| \geq l - |r'T|\right)$$

$$= MP\left(\sup_{0 \leq s \leq T-t}|x + \sigma W_s| \geq l - |r'T|\right)$$

$$\leq MP\left(\sup_{0 \leq s \leq T}|x + \sigma W_s| \geq l - |r'T|\right).$$

By Proposition 3.3.6 we know that if we define $T_a = \inf\{s > 0, W_s = a\}$, then $E(\exp(-\lambda T_a)) = \exp(-\sqrt{2\lambda}|a|)$. It infers that for any $a > 0$, and for any λ

$$P\left(\sup_{s \leq T} W_s \geq a\right) = P(T_a \leq T) \leq e^{\lambda T}E\left(e^{-\lambda T_a}\right) \leq e^{\lambda T}e^{-a\sqrt{2\lambda}}.$$

Minimising with respect to λ yields

$$P\left(\sup_{s \leq T} W_s \geq a\right) \leq \exp\left(-\frac{a^2}{T}\right),$$

and therefore

$$P\left(\sup_{s \leq T}(x + \sigma W_s) \geq a\right) \leq \exp\left(-\frac{|a - x|^2}{\sigma^2 T}\right).$$

Since $(-W_s)_{s \geq 0}$ is also a standard Brownian motion

$$P\left(\inf_{s \leq T}(x + \sigma W_s) \leq -a\right) = P\left(\sup_{s \leq T}(-x - \sigma W_s) \geq a\right) \leq \exp\left(-\frac{|a + x|^2}{\sigma^2 T}\right).$$

These two results imply that

$$P\left(\sup_{s \leq T}|x + \sigma W_s| \geq a\right) \leq \exp\left(-\frac{|a - x|^2}{\sigma^2 T}\right) + \exp\left(-\frac{|a + x|^2}{\sigma^2 T}\right)$$

and therefore

$$|u(t,x) - u_l(t,x)| \leq M\left(\exp\left(-\frac{|l - |r'T| - x|^2}{\sigma^2 T}\right) + \exp\left(-\frac{|l - |r'T| + x|^2}{\sigma^2 T}\right)\right). \quad (5.11)$$

This proves that for given t and x, $\lim_{l \to +\infty} u_l(t,x) = u(t,x)$. The convergence is even uniform in t and x as long as x remains in a compact set of $\mathbb{R}$.

Remark 5.2.1

- It can be proved that $P(\sup_{s \leq T} W_s \geq a) = 2P(W_T \geq a)$ (see Exercise 18 in Chapter 3). This would lead to a slightly better approximation than the one above.

- The fundamental advantage of the localisation method is that it can be used for pricing American options, and in that case the numerical approximation is compulsory. The estimate of the error will give us a hint to choose the domain of integration of the PDE. It is quite a crucial choice that determines how efficient our numerical procedure will be.

5.2.2 The finite difference method

Once the problem has been localised, we obtain the following system with Dirichlet boundary conditions:

$$(E) \quad \begin{cases} \dfrac{\partial u(t,x)}{\partial t} + \tilde{A}u(t,x) = 0 \text{ on } [0,T] \times \mathcal{O}_l \\[2mm] u(t,l) = u(t,-l) = 0 \text{ if } t \in [0,T] \\[2mm] u(T,x) = f(x) \text{ if } x \in \mathcal{O}_l. \end{cases}$$

The finite difference method is basically a discretisation in time and space of equation (E).

We shall start by discretising the differential operator $\tilde{A}$ on $\mathcal{O}_l$. In order to do this, a function $(f(x))_{x \in \mathcal{O}_l}$ taking values in an infinite space will be associated to a vector $(f_i)_{1 \leq i \leq N}$. We proceed as follows: we denote by (x_i) the sequence defined by $x_i = -l + 2il/(N+1)$, for $0 \leq i \leq N+1$, each f_i is somehow an approximation of $f(x_i)$. We specify boundary conditions on f_0, f_{N+1} in the Dirichlet case and f_0, f_1, f_N, f_{N+1} in the Neumann case.

We consider $h = 2l/(N+1)$ and $u_h = (u_h^i)_{1 \leq i \leq N}$ a vector in $\mathbb{R}^N$. The discretised version of the operator $\tilde{A}$ is called $\tilde{A}_h$ and the substitution runs as follows:

$$\text{replace} \quad b(x_i)\frac{\partial u(x_i)}{\partial x} \quad \text{with} \quad b(x_i)\frac{u_h^{i+1} - u_h^{i-1}}{2h}$$

and replace

$$\sigma^2(x_i)\frac{\partial^2 u(x)}{\partial x^2} \quad \text{with} \quad \sigma^2(x_i)\frac{\frac{u_h^{i+1} - u_h^i}{h} - \frac{u_h^i - u_h^{i-1}}{h}}{h} = \frac{u_h^{i+1} - 2u_h^i + u_h^{i-1}}{h^2}.$$

We obtain an operator $\tilde{A}_h$ defined on $\mathbb{R}^N$.

Remark 5.2.2 In the Black-Scholes case (after the usual logarithmic change of variables)

$$\tilde{A}^{\text{bs}-\log}u(x) = \frac{\sigma^2}{2}\frac{\partial^2 u(x)}{\partial x^2} + \left(r - \frac{\sigma^2}{2}\right)\frac{\partial u(x)}{\partial x} - ru(x),$$

is associated with

$$(\tilde{A}_h u_h)_i = \frac{\sigma^2}{2h^2}\left(u_h^{i+1} - 2u_h^i + u_h^{i-1}\right) + \left(r - \frac{\sigma^2}{2}\right)\frac{1}{2h}\left(u_h^{i+1} - u_h^{i-1}\right) - ru_h^i.$$

If we specify null Dirichlet boundary conditions, $\tilde{A}_h$ is then represented by the following matrix:

$$
\left(\left(\tilde{A}_h \right)_{ij} \right)_{1 \le i \le N,\, 1 \le j \le N}
=
\begin{pmatrix}
\beta & \gamma & 0 & \cdots & 0 & 0 \\
\alpha & \beta & \gamma & 0 & \cdots & 0 \\
0 & \alpha & \beta & \gamma & \cdots & 0 \\
0 & \vdots & \ddots & \ddots & \ddots & \vdots \\
0 & 0 & \cdots & \alpha & \beta & \gamma \\
0 & 0 & 0 & \cdots & \alpha & \beta
\end{pmatrix}
$$

where

$$
\begin{cases}
\alpha = \dfrac{\sigma^2}{2h^2} - \dfrac{1}{2h}\left(r - \dfrac{\sigma^2}{2} \right) \\[3ex]
\beta = -\dfrac{\sigma^2}{h^2} - r \\[3ex]
\gamma = \dfrac{\sigma^2}{2h^2} + \dfrac{1}{2h}\left(r - \dfrac{\sigma^2}{2} \right).
\end{cases}
$$

If we specify null Neumann conditions, it has the following form:

$$
\begin{pmatrix}
\beta + \alpha & \gamma & 0 & \cdots & 0 & 0 \\
\alpha & \beta & \gamma & 0 & \cdots & 0 \\
0 & \alpha & \beta & \gamma & \cdots & 0 \\
0 & \vdots & \ddots & \ddots & \ddots & \vdots \\
0 & 0 & \cdots & \alpha & \beta & \gamma \\
0 & 0 & 0 & \cdots & \alpha & \beta + \gamma
\end{pmatrix}.
\tag{5.12}
$$

This discretisation in space transforms (E) into an ordinary differential equation (E_h):

$$
(E_h) \quad
\begin{cases}
\dfrac{du_h(t)}{dt} + \tilde{A}_h u_h(t) = 0 & \text{if } 0 \le t \le T \\[3ex]
u_h(T) = f_h
\end{cases}
$$

where $f_h = (f_h^i)_{1 \le i \le N}$ is the vector $f_h^i = f(x_i)$.

We are now going to discretise this equation using the so-called θ-schemes. We consider $\theta \in [0, 1]$, k a time-step such that $T = Mk$ and, we approximate the solution u_h of (E_h) at time nk by $u_{h,k}$ solution of

$$
(E_{h,k}) \begin{cases}
u^M_{h,k} = f_h \\[2mm]
n \text{ decreasing, we solve for each } n: \\[2mm]
\dfrac{u^{n+1}_{h,k} - u^n_{h,k}}{k} + \theta \tilde{A}_h u^n_{h,k} + (1 - \theta) \tilde{A}_h u^{n+1}_{h,k} = 0 \\
\text{if } 0 \leq n \leq M - 1.
\end{cases}
$$

Remark 5.2.3

- When $\theta = 0$ the scheme is explicit because $u^n_{h,k}$ is computed directly from $u^{n+1}_{h,k}$. But when $\theta > 0$, we have to solve at each step a system of the form $T u^n_{h,k} = b$, with

$$
\begin{cases}
T = \left(I - \theta k \tilde{A}_h\right) \\[2mm]
b = \left(I + (1 - \theta) k \tilde{A}_h\right) u^{n+1}_{h,k}
\end{cases}
$$

where T is a tridiagonal matrix. This is obviously more complex and more time consuming. However, these schemes are often used in practice because of their good convergence properties, as we shall see shortly.

- When $\theta = 1/2$, the algorithm is called the *Crank and Nicholson* scheme. It is often used to solve systems of type (E) when $b = 0$ and σ is constant.

- When $\theta = 1$, the scheme is said to be completely implicit.

We shall now state convergence results of the solution $u_{h,k}$ of $(E_{h,k})$ towards $u(t, x)$ the solution of (E), assuming the ellipticity condition. The reader ought to refer to Raviart and Thomas (1983) for proofs. We denote by $u^k_h(t, x)$ the function

$$
\sum_{n=1}^{M} \sum_{i=1}^{N} (u^n_{h,k})_i \mathbf{1}_{]x_i - h/2, x_i + h/2]} \times \mathbf{1}_{](n-1)k, nk]}.
$$

We also call $\delta\phi$ the approximate derivative defined by

$$
(\delta\phi)(x) = \frac{1}{h} \left(\phi(x + h/2) - \phi(x - h/2) \right).
$$

Theorem 5.2.4 *We assume that b and σ are Lipschitz and that r is a non-negative continuous function. Let us recall that $\tilde{A}f(x)$ is equal to $1/2\sigma(x)^2(\partial^2 f(x)/\partial x^2) + b(x)(\partial f(x)/\partial x) - r(x)f(x)$. We assume that the operator $\tilde{A}$ is elliptic*

$$
(-\tilde{A}u, u)_{L^2(\mathcal{O}_1)} \geq \epsilon(|u|_{L^2(\mathcal{O}_1)} + |u'|_{L^2(\mathcal{O}_1)})
$$

with $\epsilon > 0$. Then:

- When $1/2 \leq \theta \leq 1$, as h, k tend to 0

$$\lim u_h^k = u \quad \text{in} \quad L^2\left([0,T] \times \mathcal{O}_l\right)$$

$$\lim \delta u_h^k = \partial u/\partial x \quad \text{in} \quad L^2\left([0,T] \times \mathcal{O}_l\right).$$

- When $0 \leq \theta < 1/2$, as h, k tend to 0, with $\lim k/h^2 = 0$, we get

$$\lim u_h^k = u \quad \text{in} \quad L^2\left([0,T] \times \mathcal{O}_l\right)$$

$$\lim \delta u_h^k = \partial u/\partial x \quad \text{in} \quad L^2\left([0,T] \times \mathcal{O}_l\right)$$

Remark 5.2.5

- In the case $0 \leq \theta < 1/2$ we say that the scheme is *conditionally convergent* because the algorithm converges only if h, k and k/h^2 tend to 0. These algorithms are rather tricky to implement numerically and therefore they are rarely used except when $\theta = 0$.

- In the case $1/2 \leq \theta \leq 1$ we say that the scheme is *unconditionally convergent* because it converges as soon as h and k tend to 0.

Finally, we shall examine in detail how we can solve problem $(E_{h,k})$ numerically. At each time-step n we are looking for a solution of $TX = G$ where

$$\begin{cases} X &= u_{h,k}^n \\[2mm] G &= \left(I + (1-\theta)k\tilde{A}_h\right) u_{h,k}^{n+1} \\[2mm] T &= I - k\theta\tilde{A}_h. \end{cases}$$

T is a tridiagonal matrix. The following algorithm, known as the Gauss method, solves the system with a number of multiplications proportional to N. Denote $X = (x_i)_{1 \leq i \leq N}$, $G = (g_i)_{1 \leq i \leq N}$ and

$$T = \begin{pmatrix} b_1 & c_1 & 0 & \cdots & & 0 & 0 \\ a_2 & b_2 & c_2 & 0 & \cdots & & 0 \\ 0 & a_3 & b_3 & c_3 & \cdots & & 0 \\ \vdots & & \ddots & \ddots & \ddots & & \vdots \\ 0 & 0 & \cdots & a_{N-1} & b_{N-1} & c_{N-1} \\ 0 & 0 & 0 & \cdots & & a_N & b_N \end{pmatrix}.$$

The algorithm runs as follows: first, we transform T into a lower triangular matrix using the Gauss method from bottom to top.

> **Upward:**
> $b'_N = b_N$
> $g'_N = g_N$
> For $1 \le i \le N - 1$, i decreasing: .
>
> $$b'_i = b_i - c_i a_{i+1}/b'_{i+1}$$
> $$g'_i = g_i - c_i g'_{i+1}/b'_{i+1}$$

We have obtained an equivalent system $T'X = G'$, where

$$T' = \begin{pmatrix} b'_1 & 0 & 0 & \cdots & 0 & 0 \\ a_2 & b'_2 & 0 & 0 & \cdots & 0 \\ 0 & a_3 & b'_3 & 0 & \cdots & 0 \\ 0 & \vdots & \ddots & \ddots & \ddots & \vdots \\ 0 & 0 & \cdots & a_{N-1} & b'_{N-1} & 0 \\ 0 & 0 & 0 & \cdots & a_N & b'_N \end{pmatrix}.$$

To conclude, we just have to compute X starting from the top of the matrix.

> **Downward:**
> $x_1 = g'_1/b'_1$
> For $2 \le i \le N$, i increasing
> $$x_i = (g'_i - a_i x_{i-1})/b'_i.$$

Remark 5.2.6 The matrix T is not necessarily invertible. However, we can prove that it is, if for any i, we have $|a_i| + |c_i| \le |b_i|$. Whenever T is not invertible, the previous algorithm does not work. In the Black-Scholes case, it is easy to check that T satisfies the preceding condition as soon as $|r - \sigma^2/2| \le \sigma^2/h$, i.e. for sufficiently small h.

5.3 American options

5.3.1 Statement of the problem

The analysis of American options in continuous time is not straightforward. In the Black- Scholes model, we obtained the following formula for the price of an American call ($f(x) = (x - K)_+$) or an American put ($f(x) = (K - x)_+$)

$$V_t = \Phi(t, S_t)$$

where

$$\Phi(t, x) = \sup_{\tau \in \mathcal{T}_{t,T}} \mathbf{E}^* \left(e^{-r(\tau-t)} f \left(x e^{(r-\sigma^2/2)(\tau-t)+\sigma(W_\tau - W_t)} \right) \right)$$

and, under $\mathbf{P}^*$, $(W_t)_{t\ge 0}$ is a standard Brownian motion and $\mathcal{T}_{t,T}$ is the set of stopping times taking values in $[t, T]$. We showed how the American call price (on a stock offering no dividend) is equal to the European call price. Nevertheless, there is no explicit formula for the put price and we require numerical methods.

The problem to be solved is a particular case of the following general problem: given a *good* function f and a diffusion $(X_t)_{t\geq 0}$ in $\mathbb{R}^n$, solution of system (5.4), compute the function

$$\Phi(t,x) = \sup_{\tau \in \mathcal{T}_{t,T}} \mathbf{E}\left(e^{-\int_t^\tau r\left(s, X_s^{t,x}\right)ds} f\left(X_\tau^{t,x}\right)\right).$$

Notice that $\Phi(t,x) \geq f(x)$ and for $t = T$ we obtain $\Phi(T,x) = f(x)$.

Remark 5.3.1 It can be proved (see Chapter 2 for the analogy with discrete time models and Chapter 4 for the Black-Scholes case) that the process

$$e^{-\int_0^t r(s,X_s)ds} \Phi(t,X_t)$$

is the smallest martingale that dominates the process $f(X_t)$ at all times.

We just stressed the fact that the European option price is the solution of a parabolic partial differential equation. As far as American options are concerned, we obtain a similar result in terms of a *parabolic system of differential inequalities* . The following theorem, stated in rather loose terms (see Remark 5.3.3), tries to explain that.

Theorem 5.3.2 *Let us assume that u is a regular solution of the following system of partial differential inequalities:*

$$\begin{cases} \dfrac{\partial u}{\partial t} + A_t u - ru \leq 0, \quad u \geq f \quad \text{in} \quad [0,T] \times \mathbb{R}^n \\[2mm] \left(\dfrac{\partial u}{\partial t} + A_t u - ru\right)(f-u) = 0 \quad \text{in} \quad [0,T] \times \mathbb{R}^n \qquad (5.13) \\[2mm] u(T,x) = f(x) \quad \text{in} \quad \mathbb{R}^n \end{cases}$$

Then

$$u(t,x) = \Phi(t,x) = \sup_{\tau \in \mathcal{T}_{t,T}} \mathbf{E}\left(e^{-\int_t^\tau r\left(s, X_s^{t,x}\right)ds} f\left(X_\tau^{t,x}\right)\right).$$

Proof. We shall only sketch the proof of this result. For a detailed demonstration, the reader ought to refer to Bensoussan and Lions (1978) (Chapter 3, Section 2) and Jaillet, Lamberton and Lapeyre (1990) (Section 3). We only consider the case $t = 0$ since the proof is very similar for arbitrary t. Let us denote by X_t^x the solution of (5.4) starting at x at time 0. Proposition 5.1.3 shows that the process

$$\begin{aligned} M_t &= e^{-\int_0^t r(s,X_s^x)ds} u(t, X_t^x) \\ &\quad - \int_0^t e^{-\int_0^s r(v,X_v^x)dv} \left(\frac{\partial u}{\partial t} + A_s u - ru\right)(s, X_s^x)ds \end{aligned}$$

is a martingale. By applying the optional sampling theorem (3.3.4) to this martingale between times 0 and τ, we get $\mathbf{E}(M_\tau) = \mathbf{E}(M_0)$, and since $\partial u/\partial t + A_s u -$

$ru \leq 0$

$$u(0, x) \geq \mathbf{E}\left(e^{-\int_0^\tau r(s, X_s^x)ds} u(\tau, X_\tau^x)\right).$$

We recall that $u(t, x) \geq f(x)$, thus $u(0, x) \geq \mathbf{E}\left(e^{-\int_0^\tau r(s, X_s^x)ds} f(X_\tau^x)\right)$. This proves that

$$u(0, x) \geq \sup_{\tau \in \mathcal{T}_{0, T}} \mathbf{E}\left(e^{-\int_0^\tau r(s, X_s^x)ds} f(X_\tau^x)\right) = F(0, x).$$

Now, we define $\tau_{\mathrm{opt}} = \inf\{0 \leq s \leq T, \ u(s, X_s^x) = f(X_s^x)\}$; we can show that τ_{opt} is a stopping time. Also, for s between 0 and τ_{opt}, we have $(\partial u / \partial t + A_s u - ru)(s, X_s^x) = 0$. The optional sampling theorem yields

$$u(0, x) = \mathbf{E}\left(e^{-\int_0^{\tau_{\mathrm{opt}}} r(s, X_s^x)ds} u(\tau_{\mathrm{opt}}, X_{\tau_{\mathrm{opt}}}^x)\right).$$

Because at time τ_{opt}, $u(\tau_{\mathrm{opt}}, X_{\tau_{\mathrm{opt}}}^x) = f(X_{\tau_{\mathrm{opt}}}^x)$, we can write

$$u(0, x) = \mathbf{E}\left(e^{-\int_0^{\tau_{\mathrm{opt}}} r(s, X_s^x)ds} f(X_{\tau_{\mathrm{opt}}}^x)\right).$$

That proves that $u(0, x) \leq F(0, x)$, and that $u(0, x) = F(0, x)$. We even *proved* that τ_{opt} is an optimal stopping time (i.e. the supremum is attained for $\tau = \tau_{\mathrm{opt}}$). □

Remark 5.3.3 The precise definition of system (5.13) is awkward because, even for a regular function f, the solution u is generally not C^2. The proper method consists in adopting a variational formulation of the problem (see Bensoussan and Lions (1978)). The proof that we have just sketched turns out to be tricky because we cannot apply the Itô formula to a solution of the previous inequality.

5.3.2 The American put in the Black-Scholes model

We are leaving the general framework to concentrate on the *pricing of the American put in the Black-Scholes model*.

We are working under the probability measure $\mathbf{P}^*$ such that the process $(W_t)_{t \geq 0}$ is a standard Brownian motion and the stock price S_t satisfies

$$dS_t = S_t(rdt + \sigma dW_t).$$

We saw in Section 5.1.3 how we can get an elliptic operator by introducing the process

$$X_t = \log(S_t) = \log(S_0) + \left(r - \frac{\sigma^2}{2}\right)t + \sigma W_t.$$

Its infinitesimal generator A is actually time-independent and

$$\tilde{A}^{\mathrm{bs-log}} = A^{\mathrm{bs-log}} - r = \frac{\sigma^2}{2}\frac{\partial^2}{\partial x^2} + \left(r - \frac{\sigma^2}{2}\right)\frac{\partial}{\partial x} - r.$$

If we consider $\phi(x) = (K - e^x)_+$, the partial differential inequality corresponding to the price of the American put is

$$
\begin{cases}
\dfrac{\partial v}{\partial t}(t, x) + \tilde{A}^{\mathrm{bs-log}} v(t, x) \leq 0 \quad \text{a.e. in} \quad [0, T] \times \mathbb{R} \\[2ex]
v(t, x) \geq \phi(x) \quad \text{a.e. in} \quad [0, T] \times \mathbb{R} \\[2ex]
(v(t, x) - \phi(x)) \left(\dfrac{\partial v}{\partial t}(t, x) + \tilde{A}^{\mathrm{bs-log}} v(t, x) \right) = 0 \quad \text{a.e. in} \quad [0, T] \times \mathbb{R} \\[2ex]
v(T, x) = \phi(x).
\end{cases}
$$

$$(5.14)$$

The following theorem states the results of existence and uniqueness of a solution to this partial differential inequality and establishes the connection with the American put price.

Theorem 5.3.4 *The inequality (5.14) has a unique continuous bounded solution $v(t, x)$ such that its partial derivatives in the distribution sense $\partial v / \partial x, \partial v / \partial t$, $\partial^2 v / \partial x^2$ are locally bounded. Moreover, this solution satisfies*

$$
v(t, \log(x)) = \Phi(t, x) = \sup_{\tau \in \mathcal{T}_{t, T}} \mathbf{E}^* \left(e^{-r(\tau - t)} f \left(x e^{(r - \sigma^2/2)(\tau - t) + \sigma(W_\tau - W_t)} \right) \right).
$$

The proof of this theorem can be found in Jaillet, Lamberton and Lapeyre (1990).

Numerical solution to this inequality

We are going to show how we can numerically solve inequality (5.14). Essentially, the method is similar to the one used in the European case. First, we localise the problem to work in the interval $\mathcal{O}_l =]-l, l[$. Then, we must impose boundary conditions at $\pm l$. Here is the inequality with Neumann boundary conditions

$$
\text{(A)} \begin{cases}
\dfrac{\partial v}{\partial t}(t, x) + \tilde{A}^{\mathrm{bs-log}} v(t, x) \leq 0 \quad \text{a.e. in} \quad [0, T] \times \mathcal{O}_l \\[2ex]
v(t, x) \geq \phi(x) \quad \text{a.e. in} \quad [0, T] \times \mathcal{O}_l \\[2ex]
(v - \phi) \left(\dfrac{\partial v}{\partial t}(t, x) + \tilde{A}^{\mathrm{bs-log}} v(t, x) \right) = 0 \quad \text{a.e. in} \quad [0, T] \times \mathcal{O}_l \\[2ex]
v(T, x) = \phi(x) \\[2ex]
\dfrac{\partial v}{\partial x}(t, \pm l) = 0.
\end{cases}
$$

We can now discretise inequality (A) using the finite differences method. The notations are the same as in Section 5.2.2. In particular, M is the integer such that

$Mk = T$, f_h is the vector given by $f_h^i = \phi(x_i)$ where $x_i = -l + 2il/(N+1)$ and $\tilde{A}_h$ is represented by matrix (5.12). If u and v are two vectors in $\mathbb{R}^n$, we write $u \leq v$ if $\forall 1 \leq i \leq n$, $u_i \leq v_i$. Formally, the method is the same as in the European case: the discretisation in time leads to the finite dimensional inequality $(A_{h,k})$:

$$(A_{h,k}) \begin{cases} u_{h,k}^M = f_h \\[1em] \text{and if } 0 \leq n \leq M - 1 \\[1em] u_{h,k}^n \geq f_h \\[1em] u_{h,k}^{n+1} - u_{h,k}^n + k \left(\theta \tilde{A}_h u_{h,k}^n + (1 - \theta) \tilde{A}_h u_{h,k}^{n+1} \right) \leq 0 \\[1em] \left(u_{h,k}^{n+1} - u_{h,k}^n + k \left(\theta \tilde{A}_h u_{h,k}^n + (1 - \theta) \tilde{A}_h u_{h,k}^{n+1} \right), u_{h,k}^n - f_h \right) = 0 \end{cases}$$

where (x, y) is the scalar product in $\mathbb{R}^N$ and $\tilde{A}_h$ is given by (5.12). If we note

$$\begin{cases} T &=& I - k\theta \tilde{A}_h \\[1em] X &=& u_{h,k}^n \\[1em] G &=& \cdot \left(I + k(1 - \theta) \tilde{A}_h \right) u_{h,k}^{n+1} \\[1em] F &=& f_h, \end{cases}$$

we have to solve, at each time n, the system of inequalities

$$(AD) \begin{cases} TX \geq G \\[1em] X \geq F \\[1em] (TX - G, X - F) = 0, \end{cases}$$

where T is the tridiagonal matrix

$$T = \begin{pmatrix} a+b & c & 0 & \cdots & 0 & 0 \\ a & b & c & 0 & \cdots & 0 \\ 0 & a & b & c & \cdots & 0 \\ 0 & \vdots & \ddots & \ddots & \ddots & \vdots \\ 0 & 0 & \cdots & a & b & c \\ 0 & 0 & 0 & \cdots & a & b+c \end{pmatrix}$$

with

$$\begin{cases} a = \theta k \left(-\dfrac{\sigma^2}{2h^2} + \dfrac{1}{2h}\left(r - \dfrac{\sigma^2}{2} \right) \right) \\[2ex] b = 1 + \theta k \left(\dfrac{\sigma^2}{h^2} + r \right) \\[2ex] c = -\theta k \left(\dfrac{\sigma^2}{2h^2} + \dfrac{1}{2h}\left(r - \dfrac{\sigma^2}{2} \right) \right), \end{cases}$$

(AD) is a finite dimensional inequality. We know how to solve this type of inequality both theoretically and numerically if the matrix T is coercive (i.e. $X.TX \geq \alpha X.X$, with $\alpha > 0$). In our case, T will satisfy this assumption if $|r - \sigma^2/2| \leq \sigma^2/h$ and if $|r - \sigma^2/2| k/2h < 1$. Indeed, this condition implies that a and c are negative and, therefore, by using the fact that $(a+b)^2 \leq 2(a^2+b^2)$ we show that

$$\begin{aligned} x.Tx &= \sum_{i=2}^{n} a x_{i-1} x_i + \sum_{i=1}^{n} b x_i^2 + \sum_{i=1}^{n-1} c x_i x_{i+1} + a x_1^2 + c x_n^2 \\ &\geq (a/2) \sum_{i=2}^{n} \left(x_{i-1}^2 + x_i^2 \right) \\ &\quad + \sum_{i=1}^{n} b x_i^2 + (c/2) \sum_{i=1}^{n-1} \left(x_i^2 + x_{i+1}^2 \right) + a x_1^2 + c x_n^2 \\ &\geq \left(a + b + c - \frac{1}{2}|a - c| \right) \sum_{i=1}^{n} x_i^2 \geq \left(1 - \frac{k}{2h}\left| r - \frac{\sigma^2}{2} \right| \right) \sum_{i=1}^{n} x_i^2. \end{aligned}$$

Under the coercitivity assumption, we can prove that there exists a unique solution to the problem $(A_{h,k})$ (see Exercise 28).

The following theorem analyses explicitly the nature of the convergence of a solution of $(A_{h,k})$ to the solution of (A). We note

$$u_h^k(t, x) = \sum_{n=1}^{M} \sum_{i=1}^{N} (u_{h,k}^n)_i \mathbb{1}_{]x_i - h/2, x_i + i/2]} \times \mathbb{1}_{](n-1)k, nk]}.$$

Theorem 5.3.5 *If u is a solution of (A),*

1. when $\theta < 1$, the convergence is conditional: if h and k converge to 0 and if k/h^2 converges to 0 then

$$\lim u_h^k = u \quad \text{in} \quad L^2\left([0,T] \times \mathcal{O}_l \right)$$

$$\lim \delta u_h^k = \frac{\partial u}{\partial x} \quad \text{in} \quad L^2\left([0,T] \times \mathcal{O}_l \right),$$

2. when $\theta = 1$, the convergence is unconditional, *i.e. the previous convergence is true when h and k converge to 0 without restriction.*

The reader will find the proof of this result in Glowinsky, Lions and Trémolières (1976). See also X.L. Zhang (1994).

Remark 5.3.6 In practice, we normally use $\theta = 1$ because the convergence is unconditional.

Numerical solution of a finite dimensional inequality

In the *American put* case, when *the step h is sufficiently small*, we can solve the system (AD) very efficiently by modifying slightly the algorithm used to solve tridiagonal systems of equations. We shall proceed as follows (we denote by b the vector $(a + b, b, \ldots, b + c)$)

> **Upward:**
> $b'_N = b_N$
> $g'_N = g_N$
> For $1 \leq i \leq N - 1$, decreasing i
> $\qquad b'_i = b_i - ca/b'_{i+1}$
> $\qquad g'_i = g_i - cg'_{i+1}/b'_{i+1}$
>
> **'American' downward:**
> $x_1 = g'_1/b'_1$
> For $2 \leq i \leq N$, increasing i
> $\qquad \tilde{x}_i = (g'_i - ax_{k-1})/b'_i$
> $\qquad x_i = \sup(\tilde{x}_i, f_i).$

Jaillet, Lamberton and Lapeyre (1990) prove that, under the previous assumptions, this algorithm does compute a solution of inequality (AD).

Remark 5.3.7 The algorithm is exactly the same as in the European case, apart from the step $x_i = \sup(\tilde{x}_i, f_i)$. That makes it very effective.

There exist other algorithms to solve inequalities in finite dimensions. Some exact methods are described in Jaillet, Lamberton and Lapeyre (1990), some iterative methods are exposed in Glowinsky, Lions and Trémolières (1976).

Remark 5.3.8 When we plug in $\theta = 1$ in $(A_{h,k})$, and we impose Neumann boundary conditions, the previous algorithm is due to Brennan and Schwartz (1977).

We must emphasise the fact that the previous algorithm only computes the exact solution of system (AD) if the assumptions stated above are satisfied. In particular, it works specifically for the American put. There exist some cases where the result computed by the previous algorithm is not the solution of (AD). The following example should erase any doubts:

$$M = \begin{pmatrix} 1 & -1 & 0 \\ -\epsilon & 1 & 0 \\ 0 & 0 & 1 \end{pmatrix}, \quad F = \begin{pmatrix} 1 \\ 2 \\ 0 \end{pmatrix}, \quad G = \begin{pmatrix} 0 \\ 0 \\ 1 \end{pmatrix}.$$

The computation gives

$$X = \begin{pmatrix} 1 \\ 2 \\ 0 \end{pmatrix},$$

which is not a solution of (AD).

Remark 5.3.9 An implementation of the *Brennan and Schwartz algorithm* is offered in Chapter 8.

5.3.3 American put pricing by a binomial method

We shall now explain another numerical method that is widely used to price the American put in the Black-Scholes model. Let r, a, b be three real numbers such that $-1 < a < r < b$. Let $(S_n)_{n\geq 0}$ be the binomial model defined by $S_0 = x$ and $S_{n+1} = S_n T_n$, where $(T_n)_{n\geq 0}$ is a sequence of IID random variables such that $\mathbf{P}(T_n = 1 + a) = p = (b - r)/(b - a)$ and $\mathbf{P}(T_n = 1 + b) = 1 - p$. We saw in Chapter 2, Exercise 4, that the American put price in this model could be written as

$$\mathcal{P}_n = P_{am}(n, S_n),$$

and that the function $P_{am}(n, x)$ could be computed by induction according to the equation

$$P_{am}(n, x) = \max \left((K - x)^+, \right.$$
$$\left. \frac{pP_{am}(n + 1, (1 + a)x) + (1 - p)P_{am}(n + 1, (1 + b)x)}{1 + r} \right)$$

$$(5.15)$$

with the final condition $P_{am}(N, x) = (K - x)^+$. On the other hand, we proved in Chapter 1, Section 1.4, that if the parameters are chosen as follows:

$$\begin{cases} r & = & RT/N \\ \\ 1 + a & = & \exp\left(-\sigma\sqrt{T/N}\right) \\ \\ 1 + b & = & \exp\left(+\sigma\sqrt{T/N}\right) \\ \\ p & = & (b - r)/(b - a), \end{cases} \qquad (5.16)$$

then the European option price in this model approximates the Black-Scholes price computed for a riskless rate equal to R and a volatility equal to σ. This suggests that in order to price the American put, we shall proceed as follow.

Given some discretisation parameter N, we fix the values r, a, b, p according to (5.16) and we compute the price $P_{am}^N(n, .)$ at the nodes $x(1 + a)^{n-i}(1 + b)^i$, $0 \leq i \leq n$ by induction of (5.15). It seems quite natural to take $P_{am}^N(0, x)$ as an approximation of the American Black-Scholes price $P(0, x)$. Indeed, we can show that $\lim_{N \to +\infty} P^N_{am}(0, x) = P(0, x)$. This result is quite tricky to justify (see Kushner (1977) and Lamberton and Pagès (1990)) and we will not try to prove it here.

This method is the so-called Cox-Ross-Rubinstein method and it is exposed in details in Cox and Rubinstein (1985).

5.4 Exercises

Exercise 28 We denote by (X, Y) the scalar product of two vectors $X = (x_i)_{1 \le i \le n}$ and $Y = (y_i)_{1 \le i \le n}$. The notation $X \ge Y$ means that for all i between 1 and n, $x_i \ge y_i$. We assume that for all X in $\mathbb{R}^n$ M satisfies $(X, MX) \ge \alpha(X, X)$ with $\alpha > 0$. We are going to study the system

$$\begin{cases} MX \ge G \\[2mm] X \ge F \\[2mm] (MX - G, X - F) = 0. \end{cases}$$

1. Show that this is equivalent to find $X \ge F$ such that

$$\forall V \ge F \quad (MX - G, V - X) \ge 0. \tag{5.17}$$

2. Prove the uniqueness of a solution of (5.17).

3. Show that if M is the identity matrix there exists a unique solution to (5.17).

4. Let ρ be positive; we denote by $S_\rho(X)$ the unique vector $Y \ge F$ such that

$$\forall V \ge F \quad (Y - X + \rho(MX - G), V - Y) \ge 0.$$

   Show that for sufficiently small ρ, S_ρ is a contraction.

5. Derive the existence of a solution to (5.17).

Exercise 29 We are trying to approximate the Black-Scholes American put price $u(t, x)$. Let us recall that u is a solution of the partial differential inequality

$$\begin{cases} \dfrac{\partial u}{\partial t}(t, x) + \tilde{A}^{bs} u(t, x) \le 0 \quad \text{a.e. in} \quad [0, T] \times]0, +\infty[\\[4mm] u(t, x) \ge (K - x)_+ \quad \text{a.e. in} \quad [0, T] \times]0, +\infty[\\[4mm] (u - (K - x)_+)\left(\dfrac{\partial u}{\partial t}(t, x) + \tilde{A}^{bs} u(t, x)\right) = 0 \quad \text{a.e. in} \quad [0, T] \times]0, +\infty[\\[4mm] u(T, x) = (K - x)_+ \end{cases}$$

where

$$\tilde{A}^{bs} = \frac{\sigma^2 x^2}{2} \frac{\partial^2}{\partial x^2} + rx \frac{\partial}{\partial x} - r.$$

1. We denote by $u_e(t, x)$ the price of the European put in the Black-Scholes model. Derive the system of inequalities satisfied by $v = u - u_e$.

2. We are going to approximate the solution $v = u - u_e$ of this inequality by discretising it in time, *using one time-step only*. When we use a totally implicit method, show that the approximation $\tilde{v}(x)$ of $v(0, x)$ satisfies

$$
\begin{cases}
-\tilde{v}(x) + T\tilde{A}^{bs}\tilde{v}(x) \leq 0 \quad \text{a.e. in} \quad]0, +\infty[\\[2mm]
\tilde{v}(t, x) \geq \tilde{\psi}(x) = (K - x)_+ - u_e(0, x) \quad \text{a.e. in} \quad]0, +\infty[\\[2mm]
\left(\tilde{v}(x) - \tilde{\psi}(x) \right) \left(-\tilde{v}(x) + T\tilde{A}^{bs}\tilde{v}(x) \right) = 0 \quad \text{a.e. in} \quad]0, +\infty[
\end{cases}
$$

$$(5.18)$$

3. Find the unique negative value for α such that $v(x) = x^\alpha$ is a solution of $-v(x) + T\tilde{A}^{bs}v(x) = 0$.

4. We look for a continuous solution of (5.18) with a continuous derivative at x^*

$$
\tilde{v}(x) =
\begin{cases}
\lambda x^\alpha \text{ if } x \geq x^* \\[2mm]
\tilde{\psi}(x) \text{ otherwise.}
\end{cases}
$$

$$(5.19)$$

Write down the equations satisfied by λ and α so that $\tilde{v}$ is continuous with continuous derivative at x^*. Deduce that if $\tilde{v}$ is continuously differentiable then x^* is a solution of $f(x) = x$ where

$$
f(x) = |\alpha| \frac{K - u_e(0, x)}{u_e'(0, x) + 1 + |\alpha|}
$$

and $u_e'(t, x) = (\partial u_e(t, x)/\partial x)$.

5. Using the closed-form formula for $u_e(0, x)$ (see Chapter 4, equation 4.9), prove that $f(0) > 0$, that $f(K) < K$ (hint: use the convexity of the function u_e) and that $f(x) - x$ is non-increasing. Conclude that there exists a unique solution to the equation $f(x) = x$.

6. Prove that $\tilde{v}(x)$ defined by (5.19) where x^* is the solution of $f(x) = x$ is a solution of (5.18).

7. Suggest an iterative algorithm (using a dichotomy argument) to compute x^* with an arbitrary accuracy.

8. From the previous results, write an algorithm in Pascal to compute the American put price.

The algorithm that we have just studied is a marginally different version of the MacMillan algorithm (see MacMillan (1986) and Barone-Adesi and Whaley (1987)).

6

Interest rate models

Interest rate models are mainly used to price and hedge bonds and bond options. Hitherto, there has not been any reference model equivalent to the Black-Scholes model for stock options. In this chapter, we will present the main features of interest rate modelling (following essentially Artzner and Delbaen (1989)), study three particular models and see how they are used in practice.

6.1 Modelling principles

6.1.1 The yield curve

In most of the models that we have already studied, the interest rate was assumed to be constant. In the real world, it is observed that the loan interest rate depends both on the date t of the loan emission and on the date T of the end or 'maturity' of the loan.

Someone borrowing one dollar at time t, until maturity T, will have to pay back an amount $F(t,T)$ at time T, which is equivalent to an average interest rate $R(t,T)$ given by the equality

$$F(t,T) = e^{(T-t)R(t,T)}.$$

If we consider the future as certain, i.e. if we assume that all interest rates $(R(t,T))_{t \leq T}$ are known, then, in an arbitrage-free world, the function F must satisfy

$$\forall t < u < s \quad F(t,s) = F(t,u)F(u,s).$$

Indeed, it is easy to derive arbitrage schemes when this equality does not hold. From this relationship and the equality $F(t,t) = 1$, it follows that, if F is smooth, there exists a function $r(t)$ such that

$$\forall t < T \quad F(t,T) = \exp\left(\int_t^T r(s)ds\right)$$

and consequently

$$R(t,T) = \frac{1}{T-t} \int_t^T r(s)ds.$$

The function $r(s)$ is interpreted as the instantaneous interest rate.

In an uncertain world, this rationale does not hold any more. At time t, the future interest rates $R(u,T)$ for $T > u > t$ are not known. Nevertheless, intuitively, it makes sense to believe that there should be some relationships between the different rates; the aim of the modelling is to determine them.

Essentially, the issue is to price bond options. We call 'zero-coupon bond' a security paying 1 dollar at a maturity date T and we note $P(t,T)$ the value of this security at time t. Obviously we have $P(T,T) = 1$ and in a world where the future is certain

$$P(t,T) = e^{-\int_t^T r(s)ds}. \tag{6.1}$$

6.1.2 Yield curve for an uncertain future

For an uncertain future, one must think of the instantaneous rate in terms of a random process: between times t and $t + dt$, it is possible to borrow at the rate $r(t)$ (in practice it corresponds to a short rate, for example the overnight rate). To make the modelling rigorous, we will consider a filtered probability space $(\Omega, \mathcal{F}, \mathbf{P}, (\mathcal{F}_t)_{0 \leq t \leq T})$ and will assume that the filtration $(\mathcal{F}_t)_{0 \leq t \leq T}$ is the natural filtration of a standard Brownian motion $(W_t)_{0 \leq t \leq T}$ and that $\mathcal{F}_T = \mathcal{F}$. As in the models we previously studied, we introduce a so-called 'riskless' asset, whose price at time t is given by

$$S_t^0 = e^{\int_0^t r(s)ds}$$

where $(r(t))_{0 \leq t \leq T}$ is an adapted process satisfying $\int_0^T |r(t)|dt < \infty$, almost surely. It might seem strange that we should call such an asset riskless since its price is random; we will see later why this asset is less 'risky' than the others. The risky assets here are the zero-coupon bonds with maturity less or equal to the horizon T. For each instant $u \leq T$, we define an adapted process $(P(t,u))_{0 \leq t \leq u}$, satisfying $P(u,u) = 1$ giving the price of the zero-coupon bond with maturity u as a function of time.

In Chapter 1, we have characterised the absence of arbitrage opportunities by the existence of an equivalent probability under which discounted asset prices are martingales. The extension of this result to continuous-time models is rather technical (cf. Harrison and Kreps (1979), Stricker (1990), Delbaen and Schacher-mayer (1994), Artzner and Delbaen (1989)), but we were able to check in Chapter 4, that such a probability exists in the Black-Scholes model. In the light of these examples, the starting point of the modelling will be based upon the following hypothesis:

(H) There is a probability $\mathbf{P}^*$ equivalent to $\mathbf{P}$, under which, for all real-valued

$u \in [0, T]$, the process $(\tilde{P}(t, u))_{0 \leq t \leq u}$ defined by

$$\tilde{P}(t, u) = e^{-\int_0^t r(s)ds} P(t, u)$$

is a martingale.

This hypothesis has some interesting consequences. Indeed, the martingale property under $\mathbf{P}^*$ leads to, using the equality $P(u, u) = 1$,

$$\tilde{P}(t, u) = \mathbf{E}^* \left(\tilde{P}(u, u) \Big| \mathcal{F}_t \right) = \mathbf{E}^* \left(e^{-\int_0^u r(s)ds} \Big| \mathcal{F}_t \right)$$

and, eliminating the discounting,

$$P(t, u) = \mathbf{E}^* \left(e^{-\int_t^u r(s)ds} \Big| \mathcal{F}_t \right). \tag{6.2}$$

This equality, which could be compared to formula (6.1), shows that the prices $P(t, u)$ only depend on the behaviour of the process $(r(s))_{0 \leq s \leq T}$ under the probability $\mathbf{P}^*$. The hypothesis we made on the filtration $(\mathcal{F}_t)_{0 \leq t \leq T}$ allows us to express the density of the probability $\mathbf{P}^*$ with respect to $\mathbf{P}$. We denote by L_T this density. For any non-negative random variable X, we have $\mathbf{E}^*(X) = \mathbf{E}(XL_T)$ and, if X is $\mathcal{F}_t$-measurable, $\mathbf{E}^*(X) = \mathbf{E}(XL_t)$, setting $L_t = \mathbf{E}(L_T | \mathcal{F}_t)$. Thus the random variable L_t is the density of $\mathbf{P}^*$ restricted to $\mathcal{F}_t$ with respect to $\mathbf{P}$.

Proposition 6.1.1 *There is an adapted process $(q(t))_{0 \leq t \leq T}$ such that, for all $t \in [0, T]$,*

$$L_t = \exp \left(\int_0^t q(s)dW_s - \frac{1}{2} \int_0^t q(s)^2 ds \right) \quad a.s. \tag{6.3}$$

Proof. The process $(L_t)_{0 \leq t \leq T}$ is a martingale relative to $(\mathcal{F}_t)$, which is the natural filtration of the Brownian motion (W_t). It follows (cf. Section 4.2.3 of Chapter 4) that there exists an adapted process $(H_t)_{0 \leq t \leq T}$ satisfying $\int_0^T H_t^2 dt < \infty$ a.s. and for all $t \in [0, T]$

$$L_t = L_0 + \int_0^t H_s dW_s \quad a.s.$$

Since L_T is a probability density, we have $\mathbf{E}(L_T) = 1 = L_0$ and, for $\mathbf{P}^*$ is equivalent to $\mathbf{P}$, we have $L_T > 0$ a.s. and more generally $\mathbf{P}(L_t > 0) = 1$ for any t. To obtain the formula (6.3), we apply the Itô formula to the log function. To do so, we need to check that $\mathbf{P}\left(\forall t \in [0, T], L_0 + \int_0^t H_s dW_s > 0\right) = 1$. The proof of this fact relies in a crucial way on the martingale property and it is the purpose of Exercise 30. Then the Itô formula yields

$$\log(L_t) = \int_0^t \frac{1}{L_s} H_s dW_s - \frac{1}{2} \int_0^t \frac{1}{L_s^2} H_s^2 ds \quad a.s.$$

which leads to equality (6.3) with $q(t) = H_t/L_t$. $\qquad \square$

Corollary 6.1.2 *The price at time t of the zero-coupon bond of maturity $u \geq t$ can be expressed as*

$$P(t,u) = \mathbf{E}\left(\exp\left(-\int_t^u r(s)ds + \int_t^u q(s)dW_s - \frac{1}{2}\int_t^u q(s)^2 ds\right)\Big| \mathcal{F}_t\right).$$
(6.4)

Proof. This follows immediately from Proposition 6.1.1 and from the following formula which is easy to derive for any non-negative random variable X:

$$\mathbf{E}^*\left(X|\mathcal{F}_t\right) = \frac{\mathbf{E}\left(XL_T|\mathcal{F}_t\right)}{L_t}.$$
(6.5)

□

The following proposition gives an economic interpretation of the process $(q(t))$ (cf. following Remark 6.1.4).

Proposition 6.1.3 *For each maturity u, there is an adapted process $(\sigma_t^u)_{0 \leq t \leq u}$ such that, on $[0, u]$,*

$$\frac{dP(t,u)}{P(t,u)} = (r(t) - \sigma_t^u q(t))dt + \sigma_t^u dW_t.$$
(6.6)

Proof. Since the process $(\tilde{P}(t,u))_{0 \leq t \leq u}$ is a martingale under $\mathbf{P}^*$, $(\tilde{P}(t,u)L_t)_{0 \leq t \leq u}$ is a martingale under $\mathbf{P}$ (see Exercise 31). Moreover, we have: $\tilde{P}(t,u)L_t > 0$ a.s., for all $t \in [0, u]$. Then, using the same rationale as in the proof of Proposition 6.1.1, we see that there exists an adapted process $(\theta_t^u)_{0 \leq t \leq u}$ such that $\int_0^u (\theta_t^u)^2 dt < \infty$ and

$$\tilde{P}(t,u)L_t = \tilde{P}(0,u)e^{\int_0^t \theta_s^u dW_s - \frac{1}{2}\int_0^t (\theta_s^u)^2 ds}.$$

Hence, using the explicit expression of L_t and getting rid of the discounting factor

$$P(t,u) = P(0,u)\exp\left(\int_0^t r(s)ds + \int_0^t (\theta_s^u - q(s))dW_s\right.$$
$$\left. -\frac{1}{2}\int_0^t ((\theta_s^u)^2 - q(s)^2)ds\right).$$

Applying the Itô formula with the exponential function, we get

$$\frac{dP(t,u)}{P(t,u)} = r(t)dt + (\theta_t^u - q(t))dW_t - \frac{1}{2}((\theta_t^u)^2 - q(t)^2)dt$$
$$+ \frac{1}{2}(\theta_t^u - q(t))^2 dt$$
$$= (r(t) + q(t)^2 - \theta_t^u q(t))dt + (\theta_t^u - q(t))dW_t,$$

which gives the equality (6.6) with $\sigma_t^u = \theta_t^u - q(t)$.

□

Remark 6.1.4 The formula (6.6) is to be related with the equality $dS_t^0 = r(t)S_t^0 dt$, satisfied by the so-called riskless asset. It is the term in dW_t which makes the bond riskier. Furthermore, the term $r(t) - \sigma_t^u q(t)$ corresponds intuitively to the average yield (i.e. in expectation) of the bond at time t (because increments of Brownian motion have zero expectation) and the term $-\sigma_t^u q(t)$ is the difference between the average yield of the bond and the riskless rate, hence the interpretation of $-q(t)$ as a 'risk premium'. Under probability $\mathbf{P}^*$, the process $(\tilde{W}_t)$ defined by $\tilde{W}_t = W_t - \int_0^t q(s)ds$ is a standard Brownian motion (Girsanov theorem), and we have

$$\frac{dP(t, u)}{P(t, u)} = r(t)dt + \sigma_t^u d\tilde{W}_t. \tag{6.7}$$

For this reason the probability $\mathbf{P}^*$ is often called the 'risk neutral' probability.

6.1.3 Bond options

To make things clearer, let us first consider a European option with maturity θ on the zero-coupon bond with maturity equal to the horizon T. If it is a call with strike price K, the value of the option at time θ is obviously $(P(\theta, T) - K)_+$ and it seems reasonable to hedge this call with a portfolio of riskless asset and zero-coupon bond with maturity T. A strategy is then defined by an adapted process $((H_t^0, H_t))_{0 \leq t \leq T}$ with values in $\mathbb{R}^2$, H_t^0 representing the quantity of riskless asset and H_t the number of bonds with maturity T held in the portfolio at time t. The value of the portfolio at time t is given by

$$V_t = H_t^0 S_t^0 + H_t P(t, T) = H_t^0 e^{\int_0^t r(s)ds} + H_t P(t, T)$$

and the self-financing condition is written, as in Chapter 4, as

$$dV_t = H_t^0 dS_t^0 + H_t dP(t, T).$$

Taking into account Proposition 6.1.3, we impose the following integrability conditions: $\int_0^T |H_t^0 r(t)| dt < \infty$ and $\int_0^T (H_t \sigma_t^u)^2 dt < \infty$ a.s. As in Chapter 4, we define admissible strategies in the following manner:

Definition 6.1.5 *A strategy* $\phi = ((H_t^0, H_t))_{0 \leq t \leq T}$ *is admissible if it is self-financing and if the discounted value* $\tilde{V}_t(\phi) = H_t^0 + H_t \tilde{P}(t, T)$ *of the corresponding portfolio is, for all t, non-negative and if* $\sup_{t \in [0, T]} \tilde{V}_t$ *is square-integrable under* $\mathbf{P}^*$.

The following proposition shows that under some assumptions, it is possible to hedge all European options with maturity $\theta < T$.

Proposition 6.1.6 *We assume* $\sup_{0 \leq t \leq T} |r(t)| < \infty$ *a.s. and* $\sigma_t^T \neq 0$ *a.s. for all* $t \in [0, \theta]$. *Let* $\theta < T$ *and let h be an* $\mathcal{F}_\theta$-*measurable random variable such that* $he^{-\int_0^\theta r(s)ds}$ *is square-integrable under* $\mathbf{P}^*$. *Then there exists an admissible strategy whose value at time* θ *is equal to h. The value at time* $t \leq \theta$ *of such a strategy is given by*

$$V_t = \mathbf{E}^* \left(e^{-\int_t^\theta r(s)ds} h \,\middle|\, \mathcal{F}_t \right).$$

Proof. The method is the same as in Chapter 4. We first observe that if $\tilde{V}_t$ is the (discounted) value at time t of an admissible strategy $((H_t^0, H_t))_{0 \leq t \leq T}$, we obtain, using the self-financing condition, the integration by parts formula and Remark 6.1.4 (cf. equation (6.7))

$$
\begin{aligned}
d\tilde{V}_t &= H_t d\tilde{P}(t, T) \\
&= H_t \tilde{P}(t, T) \sigma_t^T d\tilde{W}_t.
\end{aligned}
$$

We deduce, bearing in mind that $\sup_{t \in [0, T]} \tilde{V}_t$ is square-integrable under $\mathbf{P}^*$, that $(\tilde{V}_t)$ is a martingale under $\mathbf{P}^*$. Thus we have

$$
\forall t \leq \theta \quad \tilde{V}_t = \mathbf{E}^* \left(\tilde{V}_\theta \,\middle|\, \mathcal{F}_t \right)
$$

and, if we impose the condition $V_\theta = h$, we get

$$
V_t = e^{\int_0^t r(s)ds} \mathbf{E}^* \left(e^{-\int_0^\theta r(s)ds} h \,\middle|\, \mathcal{F}_t \right).
$$

To complete the proof, it is sufficient to find an admissible strategy having the same value at any time. To do so, one proves that there exists a process $(J_t)_{0 \leq t \leq \theta}$ such that $\int_0^\theta J_t^2 < \infty$, a.s. and

$$
he^{-\int_0^\theta r(s)ds} = \mathbf{E}^* \left(he^{-\int_0^\theta r(s)ds} \right) + \int_0^\theta J_s d\tilde{W}_s.
$$

Note that this property is not a trivial consequence of the theorem of representation of martingales because we do not know whether $he^{-\int_0^\theta r(s)ds}$ is in the σ-algebra generated by the $\tilde{W}_t$'s, $t \leq \theta$ (we only know it is in the σ-algebra $\mathcal{F}_\theta$ which can be bigger (see Exercise 32 for this particular point). Once this property is proved, it is sufficient to set

$$
H_t = \frac{J_t}{\tilde{P}(t, T)\sigma_t^T} \quad \text{and} \quad H_t^0 = \mathbf{E}^* \left(he^{-\int_0^\theta r(s)ds} \,\middle|\, \mathcal{F}_t \right) - \frac{J_t}{\sigma_t^T}
$$

for $t \leq \theta$. We check easily that $((H_t^0, H_t))_{0 \leq t \leq \theta}$ defines an admissible strategy (the hypothesis $\sup_{0 \leq t \leq T} |r(t)| < \infty$ a.s. guarantees that the condition $\int_0^\theta |r(s)H_s^0| ds < \infty$ holds) whose value at time θ is indeed equal to h. $\quad\square$

Remark 6.1.7 We have not investigated the uniqueness of the probability $\mathbf{P}^*$ and it is not clear that the risk process $(q(t))$ is defined without ambiguity. Actually, it can be shown (cf. Artzner and Delbaen (1989)) that $\mathbf{P}^*$ is the unique probability equivalent to $\mathbf{P}$ under which $(\tilde{P}(t, T))_{0 \leq t \leq T}$ is a martingale if and only if the process (σ_t^T) satisfies $\sigma_t^T \neq 0$, $dt d\mathbf{P}$ almost everywhere. This condition, slightly weaker than the hypothesis of Proposition 6.1.6, is exactly what is needed to hedge options with bonds of maturity T, which is not surprising when one keeps in mind the characterisation of complete markets we gave in Chapter 1.

6.2 Some classical models

Equations (6.2) and (6.4) show that in order to calculate the price of bonds, we need to know either the dynamics of $r(t)$ under $\mathbf{P}^*$, or the dynamics of the pair $(r(t), q(t))$ under $\mathbf{P}$. The first models we are about to examine describe the dynamics of $r(t)$ under $\mathbf{P}$ by a diffusion equation and determine the form that $q(t)$ should have to get a similar equation under $\mathbf{P}^*$. Then the prices of bonds and options depend explicitly on 'risk parameters' which are difficult to estimate. One advantage of the Heath-Jarrow-Morton model, which we will explain briefly in paragraph 6.2.3, is to provide formulae that only depend on the parameters of the dynamics of interest rates under $\mathbf{P}$.

6.2.1 The Vasicek model

In this model, we assume that the process $r(t)$ satisfies

$$dr(t) = a\left(b - r(t)\right)dt + \sigma dW_t \tag{6.8}$$

where a, b, σ are non-negative constants. We also assume that the process $q(t)$ is a constant $q(t) = -\lambda$, with $\lambda \in \mathbb{R}$. Then

$$dr(t) = a\left(b^* - r(t)\right)dt + \sigma d\tilde{W}_t \tag{6.9}$$

where $b^* = b - \lambda\sigma/a$ and $\tilde{W}_t = W_t + \lambda t$. Before calculating the price of bonds according to this model, let us give some consequences of equation (6.8). If we set

$$X_t = r(t) - b,$$

we see that (X_t) is a solution of the stochastic differential equation

$$dX_t = -aX_t dt + \sigma dW_t,$$

which means that (X_t) is an Ornstein-Uhlenbeck process (cf. Chapter 3, Section 3.5.2). We deduce that $r(t)$ can be written as

$$r(t) = r(0)e^{-at} + b\left(1 - e^{-at}\right) + \sigma e^{-at} \int_0^t e^{as} dW_t \tag{6.10}$$

and that $r(t)$ follows a normal law whose mean is given by $\mathbf{E}(r(t)) = r(0)e^{-at} + b\left(1 - e^{-at}\right)$ and variance by $\mathrm{Var}(r(t)) = \sigma^2/2a\left(1 - e^{-2at}\right)$. It follows that $\mathbf{P}(r(t) < 0) > 0$, which is not very satisfactory from a practical point of view (unless this probability is always very small). Note that, when t tends to infinity, $r(t)$ converges in law to a Gaussian random variable with mean b and variance $\sigma^2/2a$.

To calculate the price of zero-coupon bonds, we proceed under probability $\mathbf{P}^*$ and we use equation (6.9). From equality (6.2),

$$P(t, T) = \mathbf{E}^*\left(e^{-\int_t^T r(s)ds}\Big|\mathcal{F}_t\right)$$

$$= e^{-b^*(T-t)} \mathbf{E}^* \left(e^{-\int_t^T X_s^* ds} \middle| \mathcal{F}_t \right) \tag{6.11}$$

where $X_t^* = r(t) - b^*$. Since (X_t^*) is a solution of the diffusion equation with coefficients independent of time

$$dX_t = -aX_t dt + \sigma d\tilde{W}_t, \tag{6.12}$$

we can write

$$\mathbf{E}^* \left(e^{-\int_t^T X_s^* ds} \middle| \mathcal{F}_t \right) = F(T-t, X_t^*) = F(T-t, r(t) - b^*) \tag{6.13}$$

where F is the function defined by $F(\theta, x) = \mathbf{E}^* \left(e^{-\int_0^\theta X_s^x ds} \right)$, (X_t^x) being the unique solution of equation (6.12) which satisfies $X_0^x = x$ (cf. Chapter 3, Remark 3.5.11).

It is possible to calculate $F(\theta, x)$ completely. We know (cf. Chapter 3) that the process (X_t^x) is Gaussian with continuous paths. It follows that $\int_0^\theta X_s^x ds$ is a normal random variable, since the integral is the limit of Riemann sums of Gaussian components. Thus, from the expression of the Laplace transform of a Gaussian

$$\mathbf{E}^* \left(e^{-\int_0^\theta X_s^x ds} \right) = \exp \left(-\mathbf{E}^* \left(\int_0^\theta X_s^x ds \right) + \frac{1}{2} \mathrm{Var} \left(\int_0^\theta X_s^x ds \right) \right).$$

From equality $\mathbf{E}^*(X_s^x) = xe^{-as}$, we deduce

$$\mathbf{E}^* \left(\int_0^\theta X_s^x ds \right) = x \frac{1 - e^{-a\theta}}{a}.$$

For the calculation of the variance, we write

$$\mathrm{Var} \left(\int_0^\theta X_s^x ds \right) = \mathrm{Cov} \left(\int_0^\theta X_s^x ds, \int_0^\theta X_s^x ds \right)$$

$$= \int_0^\theta \int_0^\theta \mathrm{Cov} \left(X_t^x, X_u^x \right) du dt. \tag{6.14}$$

Since $X_t^x = xe^{-at} + \sigma e^{-at} \int_0^t e^{as} d\tilde{W}_s$, we have

$$\mathrm{Cov} \left(X_t^x, X_u^x \right) = \sigma^2 e^{-a(t+u)} \mathbf{E}^* \left(\int_0^t e^{as} d\tilde{W}_s \int_0^u e^{as} d\tilde{W}_s \right)$$

$$= \sigma^2 e^{-a(t+u)} \int_0^{t \wedge u} e^{2as} ds$$

$$= \sigma^2 e^{-a(t+u)} \frac{\left(e^{2a(t \wedge u)} - 1 \right)}{2a}$$

and in equality (6.14), we get

$$\mathrm{Var}\left(\int_0^\theta X_s^x ds\right) = \frac{\sigma^2 \theta}{a^2} - \frac{\sigma^2}{a^3}\left(1 - e^{-a\theta}\right) - \frac{\sigma^2}{2a^3}\left(1 - e^{-a\theta}\right)^2.$$

Going back to equations (6.11) and (6.13), we obtain the following formula

$$P(t, T) = \exp\left[-(T - t)R(T - t, r(t))\right],$$

where $R(T - t, r(t))$, which can be seen as the average interest rate on the period $[t, T]$, is given by the formula

$$R(\theta, r) = R_\infty - \frac{1}{a\theta}\left[(R_\infty - r)\left(1 - e^{-a\theta}\right) - \frac{\sigma^2}{4a^2}\left(1 - e^{-a\theta}\right)^2\right]$$

with $R_\infty = \lim_{\theta \to \infty} R(\theta, r) = b^* - \sigma^2/(2a^2)$. The yield R_∞ can be interpreted as a long-term rate; note that it does not depend on the 'instantaneous spot rate' r. This last property is considered as an imperfection of the model by practitioners.

Remark 6.2.1 In practice, parameters must be estimated and a value for r must be chosen. For r we will choose a short rate (for example, the overnight rate); then we will fit the parameters b, a, σ by statistical methods to the historical data of the instantaneous rate. Finally λ will be determined from market data by inverting the Vasicek formula. What practitioners really do is to determine the parameters, including r, by fitting the Vasicek formula on market data.

Remark 6.2.2 In the Vasicek model, the pricing of bond options is easy because of the Gaussian property of the Ornstein-Uhlenbek process (cf. Exercise 33).

6.2.2 The Cox-Ingersoll-Ross model

Cox, Ingersoll and Ross (1985) suggest modelling the behaviour of the instantaneous rate by the following equation:

$$dr(t) = (a - br(t))dt + \sigma\sqrt{r(t)}dW_t \qquad (6.15)$$

with σ and a non-negative, $b \in \mathbb{R}$, and the process $(q(t))$ being equal to $q(t) = -\alpha\sqrt{r(t)}$, with $\alpha \in \mathbb{R}$. Note that we cannot apply the theorem of existence and uniqueness that we gave in Chapter 3 because the square root function is only defined on $\mathbb{R}^+$ and is not Lipschitz. However, from the Hölder property of the square root function, one can show the following result.

Theorem 6.2.3 *We suppose that (W_t) is a standard Brownian motion defined on $[0, \infty[$. For any real number $x \geq 0$, there is a unique continuous, adapted process (X_t), taking values in $\mathbb{R}^+$, satisfying $X_0 = x$ and*

$$dX_t = (a - bX_t)\,dt + \sigma\sqrt{X_t}dW_t \quad on \ [0, \infty[. \qquad (6.16)$$

For a proof of this result, the reader is referred to Ikeda and Watanabe (1981), p. 221. To be able to study the Cox-Ingersoll-Ross model, we give some properties of this equation. We denote by (X_t^x) the solution of (6.16) starting at x and τ_0^x the stopping time defined by

$$\tau_0^x = \inf\{t \geq 0 | X_t^x = 0\}$$

with, as usual, inf $\emptyset = \infty$.

Proposition 6.2.4

1. If $a \geq \sigma^2/2$, we have $\mathbf{P}(\tau_0^x = \infty) = 1$, for all $x > 0$.
2. If $0 \leq a < \sigma^2/2$ and $b \geq 0$, we have $\mathbf{P}(\tau_0^x < \infty) = 1$, for all $x > 0$.
3. If $0 \leq a < \sigma^2/2$ and $b < 0$, we have $\mathbf{P}(\tau_0^x < \infty) \in]0, 1[$, for all $x > 0$.

This proposition is proved in Exercise 34.

The following proposition, which enables us to characterise the joint law of $\left(X_t^x, \int_0^t X_s^x ds\right)$, is the key to any pricing within the Cox-Ingersoll-Ross model.

Proposition 6.2.5 *For any non-negative λ and μ, we have*

$$\mathbf{E}\left(e^{-\lambda X_t^x} e^{-\mu \int_0^t X_s^x ds}\right) = \exp\left(-a\phi_{\lambda,\mu}(t)\right) \exp\left(-x\psi_{\lambda,\mu}(t)\right)$$

where the functions $\phi_{\lambda,\mu}$ and $\psi_{\lambda,\mu}$ are given by

$$\phi_{\lambda,\mu}(t) = -\frac{2}{\sigma^2} \log\left(\frac{2\gamma e^{\frac{t(\gamma+b)}{2}}}{\sigma^2\lambda(e^{\gamma t} - 1) + \gamma - b + e^{\gamma t}(\gamma + b)}\right)$$

and

$$\psi_{\lambda,\mu}(t) = \frac{\lambda(\gamma + b + e^{\gamma t}(\gamma - b)) + 2\mu(e^{\gamma t} - 1)}{\sigma^2\lambda(e^{\gamma t} - 1) + \gamma - b + e^{\gamma t}(\gamma + b)}$$

with $\gamma = \sqrt{b^2 + 2\sigma^2\mu}$.

Proof. The fact that this expectation can be written as $e^{-a\phi(t) - x\psi(t)}$ is due to the additivity property of the process (X_t^x) relative to the parameter a and the initial condition x (cf. Ikeda and Watanabe (1981), p. 225, Revuz and Yor (1990)). If, for λ and μ fixed, we consider the function $F(t, x)$ defined by

$$F(t, x) = \mathbf{E}\left(e^{-\lambda X_t^x} e^{-\mu \int_0^t X_s^x ds}\right), \tag{6.17}$$

it is natural to look for F as a solution of the problem

$$\begin{cases} \dfrac{\partial F}{\partial t} = \dfrac{\sigma^2}{2} x \dfrac{\partial^2 F}{\partial x^2} + (a - bx)\dfrac{\partial F}{\partial x} - \mu x F \\ \\ F(0, x) = e^{-\lambda x}. \end{cases}$$

Indeed, if F satisfies these equations and has bounded derivatives, the Itô formula shows that, for any T, the process $(M_t)_{0 \leq t \leq T}$, defined by

$$M_t = e^{-\mu \int_0^t X_s^x ds} F(T - t, X_t^x)$$

is a martingale and the equality $\mathbf{E}(M_T) = M_0$ leads to (6.17). If F can be written as $F(t, x) = e^{-a\phi(t) - x\psi(t)}$, the equations above become $\phi(0) = 0$, $\psi(0) = \lambda$ and

$$\begin{cases} -\psi'(t) &= \frac{\sigma^2}{2}\psi^2(t) + b\psi(t) - \mu \\ \phi'(t) &= \psi(t). \end{cases}$$

Solving these two differential equations gives the desired expressions for ϕ and ψ. $\qquad\square$

When applying Proposition 6.2.5 with $\mu = 0$, we obtain the Laplace transform of X_t^x

$$\begin{aligned} \mathbf{E}\left(e^{-\lambda X_t^x}\right) &= \left(\frac{b}{\sigma^2/2\lambda(1 - e^{-bt}) + b}\right)^{2a/\sigma^2} \exp\left(-x\frac{\lambda b e^{-bt}}{\sigma^2/2\lambda(1 - e^{-bt}) + b}\right) \\ &= \frac{1}{(2\lambda L + 1)^{2a/\sigma^2}} \exp\left(-\frac{\lambda L \zeta}{2\lambda L + 1}\right) \end{aligned}$$

with $L = \sigma^2/4b\left(1 - e^{-bt}\right)$ and $\zeta = 4xb/(\sigma^2\left(e^{bt} - 1\right))$. With these notations, the Laplace transform of X_t^x/L is given by the function $g_{4a/\sigma^2, \zeta}$, where $g_{\delta, \zeta}$ is defined by

$$g_{\delta, \zeta}(\lambda) = \frac{1}{(2\lambda + 1)^{\delta/2}} \exp\left(-\frac{\lambda\zeta}{2\lambda + 1}\right).$$

This function is the Laplace transform of the non-central chi-square law with δ degrees of freedom and parameter ζ (see Exercise 35 for this matter). The *density* of this law is given by the function $f_{\delta, \zeta}$, defined by

$$f_{\delta, \zeta}(x) = \frac{e^{-\zeta/2}}{2\zeta^{\delta/4 - 1/2}} e^{-x/2} x^{\delta/4 - 1/2} I_{\delta/2 - 1}\left(\sqrt{x\zeta}\right) \qquad \text{for } x > 0,$$

where I_ν is the first-order modified Bessel function with index ν, defined by

$$I_\nu(x) = \left(\frac{x}{2}\right)^\nu \sum_{n=0}^\infty \frac{(x/2)^{2n}}{n!\Gamma(\nu + n + 1)}.$$

The reader can find many properties of Bessel functions and some approximations of distribution functions of non-central chi-squared laws in Abramowitz and Stegun (1970), Chapters 9 and 26.

Let us go back to the Cox-Ingersoll-Ross model. From the hypothesis on the processes $(r(t))$ and $(q(t))$, we get

$$dr(t) = (a - (b + \sigma\alpha)r(t)) \, dt + \sigma\sqrt{r(t)}d\tilde{W}_t,$$

where, under probability $\mathbf{P}^*$, the process $(\tilde{W}_t)_{0 \le t \le T}$ is a standard Brownian motion. The price of a zero-coupon bond with maturity T is then given, at time 0, by:

$$P(0,T) = \mathbf{E}^* \left(e^{-\int_0^T r(s)ds} \right)$$

$$= e^{-a\phi(T)-r(0)\psi(T)} \tag{6.18}$$

where the functions ϕ and ψ are given by the following formulae

$$\phi(t) = -\frac{2}{\sigma^2} \log \left(\frac{2\gamma^* e^{\frac{t(\gamma^*+b^*)}{2}}}{\gamma^* - b^* + e^{\gamma^* t}(\gamma^* + b^*)} \right)$$

and

$$\psi(t) = \frac{2 \left(e^{\gamma^* t} - 1 \right)}{\gamma^* - b^* + e^{\gamma^* t}(\gamma^* + b^*)}$$

with $b^* = b + \sigma\alpha$ and $\gamma^* = \sqrt{(b^*)^2 + 2\sigma^2}$. The price at time t is given by

$$P(t,T) = \exp\left(-a\phi(T-t) - r(t)\psi(T-t)\right).$$

Let us now price a European call with maturity θ and exercise price K, on a zero-coupon bond with maturity T. We can show that the hypothesis of Proposition 6.1.6 holds; the call price at time 0 is thus given by

$$\begin{aligned}
C_0 &= \mathbf{E}^* \left[e^{-\int_0^\theta r(s)ds} \left(P(\theta,T) - K \right)_+ \right] \\
&= \mathbf{E}^* \left[e^{-\int_0^\theta r(s)ds} \left(e^{-a\phi(T-\theta)-r(\theta)\psi(T-\theta)} - K \right)_+ \right] \\
&= \mathbf{E}^* \left(e^{-\int_0^\theta r(s)ds} P(\theta,T) 1_{\{r(\theta)<r^*\}} \right) \\
&\quad - K\mathbf{E}^* \left(e^{-\int_0^\theta r(s)ds} 1_{\{r(\theta)<r^*\}} \right)
\end{aligned}$$

where r^* is defined by

$$r^* = -\frac{a\phi(T-\theta) + \log(K)}{\psi(T-\theta)}.$$

Notice that $\mathbf{E}^* \left(e^{-\int_0^\theta r(s)ds} P(\theta,T) \right) = P(0,T)$, from the martingale property of discounted prices. Similarly, $\mathbf{E}^* \left(e^{-\int_0^\theta r(s)ds} \right) = P(0,\theta)$. We can then write the price of the option as

$$C_0 = P(0,T)\mathbf{P}_1 \left(r(\theta) < r^* \right) - KP(0,\theta)\mathbf{P}_2 \left(r(\theta) < r^* \right),$$

denoting by $\mathbf{P}_1$ and $\mathbf{P}_2$ the probabilities whose densities relative to $\mathbf{P}^*$ are given respectively by

$$\frac{d\mathbf{P}_1}{d\mathbf{P}^*} = \frac{e^{-\int_0^\theta r(s)\,ds}P(\theta,T)}{P(0,T)} \quad \text{and} \quad \frac{d\mathbf{P}_2}{d\mathbf{P}^*} = \frac{e^{-\int_0^\theta r(s)\,ds}}{P(0,\theta)}.$$

We prove (cf. Exercise 36) that, if we set

$$L_1 = \frac{\sigma^2}{2} \frac{(e^{\gamma^*\theta}-1)}{\gamma^*(e^{\gamma^*\theta}+1)+(\sigma^2\psi(T-\theta)+b^*)(e^{\gamma^*\theta}-1)}$$

and

$$L_2 = \frac{\sigma^2}{2} \frac{(e^{\gamma^*\theta}-1)}{\gamma^*(e^{\gamma^*\theta}+1)+b^*(e^{\gamma^*\theta}-1)},$$

the law of $r(\theta)/L_1$ under $\mathbf{P}_1$ (resp. $r(\theta)/L_2$ under $\mathbf{P}_2$) is a non-central chi-squared law with $4a/\sigma^2$ degrees of freedom and parameter equal to ζ_1 (resp. ζ_2), with

$$\zeta_1 = \frac{8r(0)\gamma^{*2}e^{\gamma^*\theta}}{\sigma^2(e^{\gamma^*\theta}-1)(\gamma^*(e^{\gamma^*\theta}+1)+(\sigma^2\psi(T-\theta)+b^*)(e^{\gamma^*\theta}-1))}$$

and

$$\zeta_2 = \frac{8r(0)\gamma^{*2}e^{\gamma^*\theta}}{\sigma^2(e^{\gamma^*\theta}-1)(\gamma^*(e^{\gamma^*\theta}+1)+b^*(e^{\gamma^*\theta}-1))}.$$

With these notations, introducing the distribution function $F_{\delta,\zeta}$ of the non-central chi-squared law with δ degrees of freedom and parameter ζ, we have consequently

$$C_0 = P(0,T)F_{4a/\sigma^2,\zeta_1}\left(\frac{r^*}{L_1}\right) - KP(0,\theta)F_{4a/\sigma^2,\zeta_2}\left(\frac{r^*}{L_2}\right).$$

6.2.3 Other models

The main drawback of the Vasicek model and the Cox-Ingersoll-Ross model lies in the fact that prices are explicit functions of the instantaneous 'spot' interest rate so that these models are unable to take the whole yield curve observed on the market into account in the price structure.

Some authors have resorted to a two-dimensional analysis to improve the models in terms of discrepancies between short and long rates, cf. Brennan and Schwartz (1979), Schaefer and Schwartz (1984) and Courtadon (1982). These more complex models do not lead to explicit formulae and require the solution of partial differential equations. More recently, Ho and Lee (1986) have proposed a discrete-time model describing the behaviour of the whole yield curve. The continuous-time model we present now is based on the same idea and has been introduced by Heath, Jarrow and Morton (1987) and Morton (1989).

First of all we define the *forward* interest rates $f(t,s)$, for $t \le s$, characterised by the following equality:

$$P(t, u) = \exp\left(-\int_t^u f(t, s)ds\right) \tag{6.19}$$

for any maturity u. So $f(t, s)$ represents the instantaneous interest rate at time s as 'anticipated' by the market at time t. For each u, the process $(f(t, u))_{0 \le t \le u}$ must then be an adapted process and it is natural to set $f(t, t) = r(t)$. Moreover, we constrain the map $(t, s) \mapsto f(t, s)$, defined for $t \le s$, to be continuous. Then the next step of the modelling consists in assuming that, for each maturity u, the process $(f(t, u))_{0 \le t \le u}$ satisfies an equation of the following form:

$$f(t, u) = f(0, u) + \int_0^t \alpha(v, u)dv + \int_0^t \sigma(f(v, u))dW_v, \tag{6.20}$$

the process $(\alpha(t, u))_{0 \le t \le u}$ being adapted, the map $(t, u) \mapsto \alpha(t, u)$ being continuous and σ being a continuous map from $\mathbb{R}$ into $\mathbb{R}$ (σ could depend on time as well, cf. Morton (1989)).

Then we have to make sure that this model is compatible with the hypothesis (H). This gives some conditions on the coefficients α and σ of the model. To find them, we derive the differential $dP(t, u)/P(t, u)$ and we compare it to equation (6.6). Let us set $X_t = -\int_t^u f(t, s)ds$. We have $P(t, u) = e^{X_t}$ and, from equation (6.20),

$$\begin{aligned}
X_t &= \int_t^u \left(-f(s, s) + f(s, s) - f(t, s)\right) ds \\
&= -\int_t^u f(s, s)ds + \int_t^u \left(\int_t^s \alpha(v, s)dv\right) ds \\
&\quad + \int_t^u \left(\int_t^s \sigma(f(v, s))dW_v\right) ds \\
&= -\int_t^u f(s, s)ds + \int_t^u \left(\int_v^u \alpha(v, s)ds\right) dv \\
&\quad + \int_t^u \left(\int_v^u \sigma(f(v, s))ds\right) dW_v \\
&= X_0 + \int_0^t f(s, s)ds - \int_0^t \left(\int_v^u \alpha(v, s)ds\right) dv \\
&\quad - \int_0^t \left(\int_v^u \sigma(f(v, s))ds\right) dW_v. \tag{6.21}
\end{aligned}$$

The fact that the integrals commute in equation (6.21) is justified in Exercise 37. We then have

$$dX_t = \left(f(t, t) - \int_t^u \alpha(t, s)ds\right) dt - \left(\int_t^u \sigma(f(t, s))ds\right) dW_t$$

and by the Itô formula

$$\frac{dP(t,u)}{P(t,u)} = dX_t + \frac{1}{2}d\langle X, X\rangle_t$$

$$= \left(f(t,t) - \left(\int_t^u \alpha(t,s)ds\right) + \frac{1}{2}\left(\int_t^u \sigma(f(t,s))ds\right)^2\right)dt$$

$$- \left(\int_t^u \sigma(f(t,s))ds\right)dW_t.$$

If the hypothesis (H) holds, we must have, from Proposition 6.1.3 and equality $f(t,t) = r(t)$,

$$\sigma_t^u q(t) = \left(\int_t^u \alpha(t,s)ds\right) - \frac{1}{2}\left(\int_t^u \sigma(f(t,s))ds\right)^2,$$

with $\sigma_t^u = -\left(\int_t^u \sigma(f(t,s))ds\right)$. Whence

$$\int_t^u \alpha(t,s)ds = \frac{1}{2}\left(\int_t^u \sigma(f(t,s))ds\right)^2 - q(t)\int_t^u \sigma(f(t,s))ds$$

and, differentiating with respect to u,

$$\alpha(t,u) = \sigma(f(t,u))\left(\int_t^u \sigma(f(t,s))ds - q(t)\right).$$

Equation (6.20) becomes, if written in differential form,

$$df(t,u) = \sigma(f(t,u))\left(\int_t^u \sigma(f(t,s))ds\right)dt + \sigma\left(f(t,u)\right)d\tilde{W}_t. \qquad (6.22)$$

The following theorem, by Heath, Jarrow and Morton (1987), gives some sufficient conditions such that equation (6.22) has a unique solution.

Theorem 6.2.6 *If the function σ is Lipschitz and bounded, for any continuous function ϕ from $[0,T]$ to $\mathbb{R}^+$ there exists a unique continuous process with two indices $(f(t,u))_{0\leq t\leq u\leq T}$ such that, for all u, the process $(f(t,u))_{0\leq t\leq u}$ is adapted and satisfies (6.22), with $f(0,u) = \phi(u)$.*

We see that, for any continuous process $(q(t))$, it is then possible to build a model of the form (6.20): take a solution of (6.22) and set

$$\alpha(t,u) = \sigma(f(t,u))\left(\int_t^u \sigma(f(t,s))ds - q(t)\right).$$

The striking feature of this model is that the law of forward rates under $\mathbf{P}^*$ only depends on the function σ. This is a consequence of equation (6.22), in which only σ and $(\tilde{W}_t)$ appear. It follows that the price of the options only depends on the function σ. This situation is similar to Black-Scholes'. The case where σ is a constant is covered in Exercise 38. Note that the boundedness condition on σ is essential since, for $\sigma(x) = x$, there is no solution (cf. Heath, Jarrow and Morton (1987) and Morton (1989)).

Notes: To price options on bonds with coupons, the reader is referred to Jamshidian (1989) and El Karoui and Rochet (1989).

6.3 Exercises

Exercise 30 Let $(M_t)_{0 \leq t \leq T}$ be a continuous martingale such that, for any $t \in [0, T]$, $\mathbf{P}(M_t > 0) = 1$. We set

$$\tau = (\inf\{t \in [0, T] \mid M_t = 0\}) \wedge T.$$

1. Show that τ is a stopping time.

2. Using the optional sampling theorem, show that $\mathbf{E}(M_T) = \mathbf{E}(M_T 1_{\{\tau = T\}})$. Deduce that $\mathbf{P}(\{\forall t \in [0, T] \ M_t > 0\}) = 1$.

Exercise 31 Let $(\Omega, \mathcal{F}, (\mathcal{F}_t)_{0 \leq t \leq T}, \mathbf{P})$ be a filtered space and let $\mathbf{Q}$ be a probability measure absolutely continuous with respect to $\mathbf{P}$. We denote by L_t the density of the restriction of $\mathbf{Q}$ to $\mathcal{F}_t$. Let $(M_t)_{0 \leq t \leq T}$ be an adapted process. Show that $(M_t)_{0 \leq t \leq T}$ is a martingale under $\mathbf{Q}$ if and only if the process $(L_t M_t)_{0 \leq t \leq T}$ is a martingale under $\mathbf{P}$.

Exercise 32 The notations are those of Section 6.1.3. Let $(M_t)_{0 \leq t \leq T}$ be a process adapted to the filtration $(\mathcal{F}_t)$. We suppose that (M_t) is a martingale under $\mathbf{P}^*$. Using Exercise 31, show that there exists an adapted process $(H_t)_{0 \leq t \leq T}$ such that $\int_0^T H_t^2 dt < \infty$ a.s. and

$$M_t = M_0 + \int_0^t H_s d\tilde{W}_s \quad \text{a.s.}$$

for all $t \in [0, T]$.

Exercise 33 We would like to price, at time 0, a call with maturity θ and strike price K on a zero-coupon bond with maturity $T > \theta$, in the Vasicek model.

1. Show that the hypothesis of Proposition 6.1.6 does hold.

2. Show the option is exercised if and only if $r(\theta) < r^*$, where

$$r^* = R_\infty \left(1 - \frac{a(T - \theta)}{1 - e^{-a(T-\theta)}}\right) - \frac{\sigma^2 \left(1 - e^{-a(T-\theta)}\right)}{4a^2}$$

$$- \log(K) \left(\frac{a}{1 - e^{-a(T-\theta)}}\right).$$

3. Let (X, Y) be a Gaussian vector with values in $\mathbb{R}^2$ under a probability $\mathbf{P}$, and let $\tilde{\mathbf{P}}$ be a probability measure absolutely continuous with respect to $\mathbf{P}$, with density

$$\frac{d\tilde{\mathbf{P}}}{d\mathbf{P}} = \frac{e^{-\lambda X}}{\mathbf{E}(e^{-\lambda X})}.$$

Show that, under $\tilde{\mathbf{P}}$, Y is normal and give its mean and variance.

4. Using the previous question, show that under the probabilities whose densities are respectively $\exp\left(-\int_0^\theta r(s)ds\right)/P(0,\theta)$ and $\exp\left(-\int_0^T r(s)ds\right)/P(0,T)$ with respect to $\mathbf{P}^*$, the random variable $r(\theta)$ is normal. Deduce an expression for the price of the option in the form $C_0 = P(0,T)p_1 - KP(0,\theta)p_2$, for some parameters p_1 and p_2 to be calculated.

Exercise 34 The aim of this exercise is to prove Proposition 6.2.4. For x, $M > 0$, we note τ_M^x the stopping time defined by $\tau_M^x = \inf\{t \geq 0 \mid X_t^x = M\}$.

1. Let s be the function defined on $]0,\infty[$ by

$$s(x) = \int_1^x e^{2by/\sigma^2} y^{-2a/\sigma^2} dy.$$

Prove that s satisfies

$$\frac{\sigma^2}{2} x \frac{d^2 s}{dx^2} + (a - bx)\frac{ds}{dx} = 0.$$

2. For $\varepsilon < x < M$, we set $\tau_{\varepsilon,M}^x = \tau_\varepsilon^x \wedge \tau_M^x$. Show that, for any $t > 0$, we have

$$s\left(X_{t\wedge\tau_{\varepsilon,M}^x}^x\right) = s(x) + \int_0^{t\wedge\tau_{\varepsilon,M}^x} s'(X_s^x)\sigma\sqrt{X_s^x}dW_s.$$

Deduce, taking the variance on both sides and using the fact that s' is bounded from below on the interval $[\varepsilon, M]$, that $\mathbf{E}\left(\tau_{\varepsilon,M}^x\right) < \infty$, which implies that $\tau_{\varepsilon,M}^x$ is finite a.s.

3. Show that if $\varepsilon < x < M$, $s(x) = s(\varepsilon)\mathbf{P}\left(\tau_\varepsilon^x < \tau_M^x\right) + s(M)\mathbf{P}\left(\tau_\varepsilon^x > \tau_M^x\right)$.

4. We assume $a \geq \sigma^2/2$. Then prove that $\lim_{x\to 0} s(x) = -\infty$. Deduce that $\mathbf{P}\left(\tau_0^x < \tau_M^x\right) = 0$ for all $M > 0$, then that $\mathbf{P}\left(\tau_0^x < \infty\right) = 0$.

5. We now assume that $0 \leq a < \sigma^2/2$ and we set $s(0) = \lim_{x\to 0} s(x)$. Show that, for all $M > x$, we have $s(x) = s(0)\mathbf{P}\left(\tau_0^x < \tau_M^x\right) + s(M)\mathbf{P}\left(\tau_0^x > \tau_M^x\right)$ and complete the proof of Proposition 6.2.4.

Exercise 35 Let d be an integer and let $X_1, X_2, \ldots, X_d$, d be independent Gaussian random variables with unit variance and respective means $m_1, m_2, \ldots, m_d$. Show that the random variable $X = \sum_{i=1}^d X_i^2$ follows a non-central chi-squared law with d degrees of freedom and parameter $\zeta = \sum_{i=1}^d m_i^2$.

Exercise 36 Using Proposition 6.2.5, derive, for the Cox-Ingersoll-Ross model, the law of $r(\theta)$ under the probabilities $\mathbf{P}_1$ and $\mathbf{P}_2$ introduced at the end of Section 6.2.2.

Exercise 37 Let $(\Omega, \mathcal{F}, (\mathcal{F}_t)_{0\leq t\leq T}, \mathbf{P})$ be a filtered space and let $(W_t)_{0\leq t\leq T}$ be a standard Brownian motion with respect to $(\mathcal{F}_t)$. We consider a process with two indices $(H(t,s))_{0\leq t,s\leq T}$ satisfying the following properties: for any ω, the map $(t,s) \mapsto H(t,s)(\omega)$ is continuous and for any $s \in [0,T]$, the process

$(H(t, s))_{0 \le t \le T}$ is adapted. We would like to justify the equality

$$\int_0^T \left(\int_0^T H(t, s)dW_t \right) ds = \int_0^T \left(\int_0^T H(t, s)ds \right) dW_t.$$

For simplicity, we assume that $\int_0^T \mathbf{E} \left(\int_0^T H^2(t, s)dt \right) ds < \infty$ (which is sufficient to justify equality (6.21)).

1. Prove that

$$\int_0^T \mathbf{E} \left(\left| \int_0^T H(t, s)dW_t \right| \right) ds \le \int_0^T \left[\mathbf{E} \left(\int_0^T H^2(t, s)dt \right) \right]^{1/2} ds.$$

Deduce that the integral $\int_0^T \left(\int_0^T H(t, s)dW_t \right) ds$ exists.

2. Let $0 = t_0 < t_1 < \cdots < t_N = T$ be a partition of interval $[0, T]$. Remark that

$$\int_0^T \left(\sum_{i=0}^{N-1} H(t_i, s) \left(W_{t_{i+1}} - W_{t_i} \right) \right) ds$$

$$= \sum_{i=0}^{N-1} \left(\int_0^T H(t_i, s)ds \right) \left(W_{t_{i+1}} - W_{t_i} \right)$$

and justify why we can take the limit to obtain the desired equality.

Exercise 38 In the Heath-Jarrow-Morton model, we assume that the function σ is a positive constant. We would like to price a call with maturity θ and strike price K, on a zero-coupon bond with maturity $T > \theta$.

1. Show that the hypothesis of Proposition 6.1.6 holds.

2. Show that the solution of equation (6.22) is given by $f(t, u) = f(0, u) + \sigma^2 t (u - t/2) + \sigma \tilde{W}_t$. Deduce that

$$P(\theta, T) = \frac{P(0, T)}{P(0, \theta)} \exp \left(-\sigma(T - \theta)\tilde{W}_\theta - \frac{\sigma^2 \theta T(T - \theta)}{2} \right).$$

3. Derive, for $\lambda \in \mathbb{R}$, $\mathbf{E}^* \left(e^{-\sigma \int_0^\theta \tilde{W}_s ds} e^{\lambda \tilde{W}_\theta} \right)$. Deduce the law of $\tilde{W}_\theta$ under the probability measures $\mathbf{P}_1$ and $\mathbf{P}_2$ with densities with respect to $\mathbf{P}^*$ respectively given by

$$\frac{d\mathbf{P}_1}{d\mathbf{P}^*} = \frac{e^{-\int_0^\theta r(s)ds} P(\theta, T)}{P(0, T)} \quad \text{and} \quad \frac{d\mathbf{P}_2}{d\mathbf{P}^*} = \frac{e^{-\int_0^\theta r(s)ds}}{P(0, \theta)}.$$

4. Show that the price of a call at time 0 is given by

$$C_0 = P(0, T)N(d) - KP(0, \theta)N \left(d - \sigma\sqrt{\theta}(T - \theta) \right),$$

where N is the standard normal distribution function and

$$d = \frac{\sigma\sqrt{\theta}(T-\theta)}{2} - \frac{\log\left(KP(0,\theta)/P(0,T)\right)}{\sigma\sqrt{\theta}(T-\theta)}.$$

7

Asset models with jumps

In the Black-Scholes model, the share price is a continuous function of time and this property is one of the characteristics of the model. But some rare events (release of an unexpected economic figure, major political changes or even a natural disaster in a major economy) can lead to brusque variations in prices. To model this kind of phenomena, we have to introduce discontinuous stochastic processes.

Most of these models 'with jumps' have a striking feature that distinguishes them from the Black-Scholes model: they are incomplete market models, and there is no perfect hedging of options in this case. It is no longer possible to price options using a replicating portfolio. A possible approach to pricing and hedging consists in defining a notion of risk and choosing a price and a hedge in order to minimise this risk.

In this chapter, we will study the simplest models with jumps. The description of these models requires a review of the main properties of the Poisson process; this is the objective of the first section.

7.1 Poisson process

Definition 7.1.1 *Let* $(T_i)_{i \geq 1}$ *be a sequence of independent, identically exponentially distributed random variables with parameter* λ*, i.e. their density is equal to* $1_{\{x>0\}} \lambda e^{-\lambda x}$*. We set* $\tau_n = \sum_{i=1}^{n} T_i$*. We call* Poisson process with intensity λ *the process* N_t *defined by*

$$N_t = \sum_{n \geq 1} 1_{\{\tau_n \leq t\}} = \sum_{n \geq 1} n 1_{\{\tau_n \leq t < \tau_{n+1}\}}.$$

Remark 7.1.2 N_t represents the number of points of the sequence $(\tau_n)_{n \geq 1}$ which are smaller than or equal to t. We have

$$\tau_n = \inf\{t \geq 0,\ N_t = n\}.$$

The following proposition gives an explicit expression for the law of N_t for a given t.

Proposition 7.1.3 *If $(N_t)_{t\geq 0}$ is a Poisson process with intensity λ then, for any $t > 0$, the random variable N_t follows a Poisson law with parameter λ*

$$\mathbf{P}(N_t = n) = e^{-\lambda t}\frac{(\lambda t)^n}{n!}.$$

In particular we have

$$\mathbf{E}(N_t) = \lambda t, \quad \text{Var}(N_t) = \mathbf{E}(N_t^2) - (\mathbf{E}(N_t))^2 = \lambda t.$$

Moreover, for $s > 0$

$$\mathbf{E}\left(s^{N_t}\right) = \exp\left\{\lambda t (s - 1)\right\}.$$

Proof. First we notice that the law of τ_n is

$$1_{\{x>0\}}\lambda e^{-\lambda x}\frac{(\lambda x)^{n-1}}{(n-1)!}dx,$$

i.e. a gamma law with parameters λ and n. Indeed, the Laplace transform of T_1 is

$$\mathbf{E}\left(e^{-\alpha T_1}\right) = \frac{\lambda}{\lambda + \alpha},$$

thus the law of $\tau_n = T_1 + \cdots + T_n$ is

$$\mathbf{E}\left(e^{-\alpha \tau_n}\right) = \mathbf{E}\left(e^{-\alpha T_1}\right)^n = \left(\frac{\lambda}{\lambda + \alpha}\right)^n.$$

We recognise the Laplace transform of the gamma law with parameters λ and n (cf. Bouleau (1986), Chapter VI, Section 7.12). Then we have, for $n \geq 1$

$$
\begin{aligned}
\mathbf{P}\left(N_t = n\right) &= \mathbf{P}\left(\tau_n \leq t\right) - \mathbf{P}\left(\tau_{n+1} \leq t\right) \\
&= \int_0^t \lambda e^{-\lambda x}\frac{(\lambda x)^{n-1}}{(n-1)!}dx - \int_0^t \lambda e^{-\lambda x}\frac{(\lambda x)^n}{n!}dx \\
&= \frac{(\lambda t)^n}{n!}e^{-\lambda t}.
\end{aligned}
$$

$\square$

Proposition 7.1.4 *Let $(N_t)_{t\geq 0}$ be a Poisson process with intensity λ and $\mathcal{F}_t = \sigma(N_s, \; s \leq t)$. The process $(N_t)_{t\geq 0}$ is a process with independent and stationary increments, i.e.*

- independence: *if $s > 0$, $N_{t+s} - N_t$ is independent of the σ-algebra $\mathcal{F}_t$.*
- stationarity: *the law of $N_{t+s} - N_t$ is identical to the law of $N_s - N_0 = N_s$.*

Remark 7.1.5 It is easy to see that the jump times τ_n are stopping times. Indeed, $\{\tau_n \leq t\} = \{N_t \geq n\} \in \mathcal{F}_t$. A random variable T with exponential law satisfies $\mathbf{P}(T \geq t + s | T \geq t) = \mathbf{P}(T \geq s)$. The exponential variables are said to be 'memoryless'. The independence of the increments is a consequence of this property of exponential laws.

Remark 7.1.6 The law of a Poisson process with intensity λ is characterised by either of the following two properties:

- $(N_t)_{t \geq 0}$ is a right-continuous homogeneous Markov process with left-hand limit, such that

$$\mathbf{P}(N_t = n) = e^{-\lambda t} \frac{(\lambda t)^n}{n!}.$$

- $(N_t)_{t \geq 0}$ is a process with independent and stationary increments, right-continuous, non-decreasing, the amplitude of the jumps being one.

For the first characterisation, cf. Bouleau (1988), Chapter III; for the second one, cf. Dacunha-Castelle and Duflo (1986), Section 6.3.

7.2 Dynamics of the risky asset

The objective of this section is to model a financial market in which there is one riskless asset (with price $S_t^0 = e^{rt}$, at time t) and one risky asset whose price jumps in the proportions $U_1, \ldots, U_j, \ldots$, at some times $\tau_1, \ldots, \tau_j, \ldots$ and which, between two jumps, follows the Black-Scholes model. Moreover, we will assume that the τ_j's correspond to the jump times of a Poisson process. To be more rigorous, let us consider a probability space $(\Omega, \mathcal{A}, \mathbf{P})$ on which we define a standard Brownian motion $(W_t)_{t \geq 0}$, a Poisson process $(N_t)_{t \geq 0}$ with intensity λ and a sequence $(U_j)_{j \geq 1}$ of independent, identically distributed random variables taking values in $]-1, +\infty[$. We will assume that the σ-algebras generated respectively by $(W_t)_{t \geq 0}$, $(N_t)_{t \geq 0}$, $(U_j)_{j \geq 1}$ are independent.

For all $t \geq 0$, let us denote by $\mathcal{F}_t$ the σ-algebra generated by the random variables W_s, N_s for $s \leq t$ and $U_j 1_{\{j \leq N_t\}}$ for $j \geq 1$. It can be shown that $(W_t)_{t \geq 0}$ is a standard Brownian motion with respect to the filtration $(\mathcal{F}_t)_{t \geq 0}$, that $(N_t)_{t \geq 0}$ is a process adapted to this filtration and that, for all $t > s$, $N_t - N_s$ is independent of the σ-algebra $\mathcal{F}_s$. Because the random variables $U_j 1_{\{j \leq N_t\}}$ are $\mathcal{F}_t$-measurable, we deduce that, at time t, the relative amplitudes of the jumps taking place before t are known. Note as well that the τ_j's are stopping times of $(\mathcal{F}_t)_{t \geq 0}$, since $\{\tau_j \leq t\} = \{N_t \geq j\} \in \mathcal{F}_t$.

The dynamics of X_t, price of the risky asset at time t, can now be described in the following manner. The process $(X_t)_{t \geq 0}$ is an adapted, right-continuous process satisfying:

- On the time intervals $[\tau_j, \tau_{j+1}[$

$$dX_t = X_t(\mu dt + \sigma dW_t).$$

- At time τ_j, the jump of X_t is given by

$$\Delta X_{\tau_j} = X_{\tau_j} - X_{\tau_j^-} = X_{\tau_j^-} U_j,$$

thus $X_{\tau_j} = X_{\tau_j^-}(1 + U_j)$.

So we have, for $t \in [0, \tau_1[$

$$X_t = X_0 e^{(\mu - \sigma^2/2)t + \sigma W_t},$$

consequently, the left-hand limit at τ_1 is given by

$$X_{\tau_1^-} = X_0 e^{(\mu - \sigma^2/2)\tau_1 + \sigma W_{\tau_1}}$$

and

$$X_{\tau_1} = X_0 (1 + U_1) e^{(\mu - \sigma^2/2)\tau_1 + \sigma W_{\tau_1}}.$$

Then, for $t \in [\tau_1, \tau_2[$,

$$
\begin{aligned}
X_t &= X_{\tau_1} e^{(\mu - \sigma^2/2)(t - \tau_1) + \sigma(W_t - W_{\tau_1})} \\
&= X_{\tau_1^-}(1 + U_1) e^{(\mu - \sigma^2/2)(t - \tau_1) + \sigma(W_t - W_{\tau_1})} \\
&= X_0 (1 + U_1) e^{(\mu - \sigma^2/2)t + \sigma W_t}.
\end{aligned}
$$

Repeating this scheme, we obtain

$$X_t = X_0 \left(\prod_{j=1}^{N_t} (1 + U_j) \right) e^{(\mu - \sigma^2/2)t + \sigma W_t},$$

with the convention $\prod_{j=1}^0 = 1$.

The process $(X_t)_{t \geq 0}$ is obviously right-continuous, adapted and has only finitely many discontinuities on each interval $[0, t]$. We can also prove that it satisfies, for all $t \geq 0$,

$$\mathbf{P} \text{ a.s.} \quad X_t = X_0 + \int_0^t X_s \left(\mu ds + \sigma dW_s \right) + \sum_{j=1}^{N_t} X_{\tau_j^-} U_j. \qquad (7.1)$$

We will see that, for this kind of model, it is generally impossible to hedge the options perfectly. This difficulty is due to the fact that for $T < +\infty$, there are infinitely many probabilities equivalent to $\mathbf{P}$ on $\mathcal{F}_T$ under which the discounted price $(e^{-rt} X_t)_{0 \leq t \leq T}$ is a martingale. In the remainder, we will make the following *assumption*: under $\mathbf{P}$, the process $(e^{-rt} X_t)_{0 \leq t \leq T}$ is a martingale. This is a stringent hypothesis, but it will allow us to determine simply some hedging strategies with minimal risk. When this hypothesis does not hold, the hedging of options is rather tricky (see Schweizer (1989)).

To derive $\mathbf{E}(X_t | \mathcal{F}_s)$ we will need the following lemma; which means intuitively that the relative amplitudes of the jumps which take place after time s are independent of the σ-algebra $\mathcal{F}_s$.

Lemma 7.2.1 *For all $s \geq 0$, the σ-algebras*

$$\sigma \left(U_{N_s+1}, U_{N_s+2}, \ldots, U_{N_s+k}, \ldots \right)$$

and $\mathcal{F}_s$ are independent.

Proof. As the σ-algebras $\mathcal{W} = \sigma(W_s, \ s \geq 0), \mathcal{N} = \sigma(N_s, \ s \geq 0)$ and $\mathcal{U} = \sigma(U_i, \ i \geq 1)$ are independent, it suffices to prove that the σ-algebra $\sigma \left(U_{N_s+1}, U_{N_s+2}, \ldots, U_{N_s+k}, \ldots \right)$

is independent of the σ-algebra generated by the random variables N_u, $u \leq s$ and $U_j 1_{\{j \leq N_s\}}$. Let A be a Borel subset of $\mathbb{R}^k$, B a Borel subset of $\mathbb{R}^d$ and C an event of the σ-algebra $\sigma(N_u,\ u \leq s)$. We have, using the independence of $\mathcal{U}$ and $\mathcal{N}$ and the fact that the U_j's are independent and identically distributed,

$$\mathbf{P}\left(\{(U_{N_s+1},\ldots,U_{N_s+k}) \in A\} \cap C \cap \{(U_1,\ldots,U_d) \in B\} \cap \{d \leq N_s\}\right)$$

$$= \sum_{p=d}^{\infty} \mathbf{P}\left(\{(U_{p+1},\ldots,U_{p+k}) \in A\} \cap \{(U_1,\ldots,U_d) \in B\} \cap C \cap \{N_s = p\}\right)$$

$$= \sum_{p=d}^{\infty} \mathbf{P}\left((U_{p+1},\ldots,U_{p+k}) \in A\right) \mathbf{P}\left((U_1,\ldots,U_d) \in B\right) \mathbf{P}\left(C \cap \{N_s = p\}\right)$$

$$= \mathbf{P}\left((U_1,\ldots,U_k) \in A\right) \sum_{p=d}^{\infty} \mathbf{P}\left((U_1,\ldots,U_d) \in B\right) \mathbf{P}\left(C \cap \{N_s = p\}\right).$$

From the last equality, we deduce (taking $C = \Omega$ and $B = \mathbb{R}^d$) that the vector $(U_{N_s+1},\ldots,U_{N_s+k})$ follows the same law as $(U_1,\ldots,U_k)$, and then that

$$\mathbf{P}\left(\{(U_{N_s+1},\ldots,U_{N_s+k}) \in A\} \cap C \cap \{(U_1,\ldots,U_d) \in B\} \cap \{d \leq N_s\}\right)$$

$$= \mathbf{P}\left((U_{N_s+1},\ldots,U_{N_s+k} \in A\right) \mathbf{P}\left(C \cap \{(U_1,\ldots,U_d \in B)\} \cap \{d \leq N_s\}\right).$$

Whence the independence stated above. $\qquad\square$

Suppose now that $\mathbf{E}(|U_1|) < +\infty$ and set $\tilde{X}_t = e^{-rt}X_t$. Then

$$\mathbf{E}(\tilde{X}_t|\mathcal{F}_s) = \tilde{X}_s \mathbf{E}\left(e^{(\mu-r-\sigma^2/2)(t-s)+\sigma(W_t-W_s)} \prod_{j=N_s+1}^{N_t} (1+U_j)\Big|\mathcal{F}_s\right)$$

$$= \tilde{X}_s \mathbf{E}\left(e^{(\mu-r-\sigma^2/2)(t-s)+\sigma(W_t-W_s)} \prod_{j=1}^{N_t-N_s} (1+U_{N_s+j})\Big|\mathcal{F}_s\right)$$

$$= \tilde{X}_s \mathbf{E}\left(e^{(\mu-r-\sigma^2/2)(t-s)+\sigma(W_t-W_s)} \prod_{j=1}^{N_t-N_s} (1+U_{N_s+j})\right),$$

using Lemma 7.2.1 and the fact that $W_t - W_s$ and $N_t - N_s$ are independent of the σ-algebra $\mathcal{F}_s$. Hence

$$\mathbf{E}(\tilde{X}_t|\mathcal{F}_s) = \tilde{X}_s e^{(\mu-r)(t-s)} \mathbf{E}\left(\prod_{j=N_s+1}^{N_t} (1+U_j)\right)$$

$$= \tilde{X}_s e^{(\mu-r)(t-s)} e^{\lambda(t-s)\mathbf{E}(U_1)},$$

using Exercise 39.

It is now clear that $(\tilde{X}_t)$ is a martingale if and only if

$$\mu = r - \lambda \mathbf{E}(U_1).$$

To deal with the terms due to the jumps in the hedging schemes, we will need two more lemmas, whose proofs can be omitted at first reading. We will denote by ν the common law of the random variables U_j's.

Lemma 7.2.2 *Let $\Phi(y, z)$ be a measurable function from $\mathbb{R}^d \times \mathbb{R}$ to $\mathbb{R}$, such that for any real number z the function $y \mapsto \Phi(y, z)$ is continuous on $\mathbb{R}^d$, and let $(Y_t)_{t \geq 0}$ be a left-continuous process, taking values in $\mathbb{R}^d$, adapted to the filtration $(\mathcal{F}_t)_{t \geq 0}$. We assume that, for all $t > 0$,*

$$\mathbf{E} \left(\int_0^t ds \int \nu(dz) \Phi^2(Y_s, z) \right) < +\infty.$$

Then the process M_t defined by

$$M_t = \sum_{j=1}^{N_t} \Phi(Y_{\tau_j}, U_j) - \lambda \int_0^t ds \int \nu(dz) \Phi(Y_s, z),$$

is a square-integrable martingale and

$$M_t^2 - \lambda \int_0^t ds \int \nu(dz) \Phi^2(Y_s, z)$$

is a martingale.

Notice that by convention $\sum_{j=1}^0 = 1$.

Proof. We assume first that Φ is bounded and we set

$$C = \sup_{(y,z) \in \mathbb{R}^d \times \mathbb{R}} |\Phi(y, z)|.$$

Then we have $\left| \sum_{j=1}^{N_t} \Phi(Y_{\tau_j}, U_j) \right| \leq C N_t$ and $\left| \int_0^t \int \nu(dz) \Phi(Y_s, z) \right| \leq Ct$. So M_t is square-integrable. Let us fix s and t, with $s < t$, and set

$$Z = \sum_{j=N_s+1}^{N_t} \Phi(Y_{\tau_j}, U_j).$$

To a partition $\rho = (s_0 = s < s_1 < \cdots < s_m = t)$ of the interval $[s, t]$, let us associate

$$Z^\rho = \sum_{i=0}^{m-1} \sum_{j=N_{s_i}+1}^{N_{s_{i+1}}} \Phi(Y_{s_i}, U_j).$$

The left-continuity of $(Y_t)_{t \geq 0}$ and the continuity of Φ with respect to y imply that Z^ρ converges almost surely to Z when the mesh of the partition ρ tends to 0.

Moreover $|Z^\rho| \le C(N_t - N_s)$; it follows that the convergence takes place in L^1 and even in L^2.

We have

$$\mathbf{E}(Z^\rho | \mathcal{F}_s) = \mathbf{E}\left(\sum_{i=0}^{m-1} \mathbf{E}(Z_{i+1} | \mathcal{F}_{s_i}) \Big| \mathcal{F}_s \right) \qquad (7.2)$$

setting

$$Z_{i+1} = \sum_{j=N_{s_i}+1}^{N_{s_{i+1}}} \Phi(Y_{s_i}, U_j) = \sum_{j=1}^{N_{s_{i+1}}-N_{s_i}} \Phi(Y_{s_i}, U_{N_{s_i}+j}).$$

Using Lemma 7.2.1 and the fact that Y_{s_i} is $\mathcal{F}_{s_i}$-measurable, we apply Proposition A.2.5 of the Appendix to see that

$$\mathbf{E}(Z_{i+1} | \mathcal{F}_{s_i}) = \bar{\Phi}_i(Y_{s_i}),$$

where $\bar{\Phi}_i(y)$ is defined by

$$\bar{\Phi}_i(y) = \mathbf{E}\left(\sum_{j=1}^{N_{s_{i+1}}-N_{s_i}} \Phi(y, U_{N_{s_i}+j}) \right).$$

$\bar{\Phi}_i(y)$ is thus the expectation of a random sum and, from Exercise 40,

$$\bar{\Phi}_i(y) = \lambda(s_{i+1} - s_i) \int d\nu(z) \Phi(y, z).$$

Going back to equation (7.2), we deduce

$$\mathbf{E}(Z^\rho | \mathcal{F}_s) = \mathbf{E}\left(\sum_{i=0}^{m-1} \bar{\Phi}_i(Y_{s_i}) \Big| \mathcal{F}_s \right) = \mathbf{E}\left(\sum_{i=0}^{m-1} \lambda(s_{i+1} - s_i) \int d\nu(z) \Phi(Y_{s_i}, z) \Big| \mathcal{F}_s \right).$$

When the mesh of ρ tends to 0, we obtain

$$\mathbf{E}\left(\sum_{j=N_s+1}^{N_t} \Phi(Y_{\tau_j}, U_j) \Big| \mathcal{F}_s \right) = \mathbf{E}\left(\lambda \int_s^t du \int d\nu(z) \Phi(Y_u, z) \Big| \mathcal{F}_s \right),$$

which proves that M_t is a martingale. Now set $\bar{Z}^\rho = \sum_{i=0}^{m-1} \mathbf{E}(Z_{i+1} | \mathcal{F}_{s_i})$. We can write

$$\bar{Z}^\rho = \sum_{i=0}^{m-1} \bar{\Phi}_i(Y_{s_i}) = \sum_{i=0}^{m-1} \lambda(s_{i+1} - s_i) \int d\nu(z) \Phi(Y_{s_i}, z).$$

Moreover,

$$
\begin{aligned}
& \mathbf{E}\left((Z^\rho - \bar{Z}^\rho)^2 \big| \mathcal{F}_s\right) \\
& = \mathbf{E}\left[\left(\sum_{i=0}^{m-1} [Z_{i+1} - \mathbf{E}(Z_{i+1}|\mathcal{F}_{s_i})]\right)^2 \bigg| \mathcal{F}_s\right] \\
& = \mathbf{E}\left(\sum_{i=0}^{m-1} [Z_{i+1} - \mathbf{E}(Z_{i+1}|\mathcal{F}_{s_i})]^2 \bigg| \mathcal{F}_s\right) \\
& \quad + 2\sum_{i<j} \mathbf{E}\left((Z_{i+1} - \mathbf{E}(Z_{i+1}|\mathcal{F}_{s_i}))(Z_{j+1} - \mathbf{E}(Z_{j+1}|\mathcal{F}_{s_j})) \bigg| \mathcal{F}_s\right).
\end{aligned}
$$

Taking the conditional expectation with respect to $\mathcal{F}_{s_j}$ and using the fact that Z_{i+1} is $\mathcal{F}_{s_{i+1}}$ hence $\mathcal{F}_{s_j}$-measurable, we see that the second sum is 2. Whence

$$
\begin{aligned}
\mathbf{E}\left((Z^\rho - \bar{Z}^\rho)^2 \big| \mathcal{F}_s\right) & = \mathbf{E}\left(\sum_{i=0}^{m-1} (Z_{i+1} - \mathbf{E}(Z_{i+1}|\mathcal{F}_{s_i}))^2 \bigg| \mathcal{F}_s\right) \\
& = \mathbf{E}\left(\sum_{i=0}^{m-1} \mathbf{E}\left([Z_{i+1} - \mathbf{E}(Z_{i+1}|\mathcal{F}_{s_i})]^2 \big| \mathcal{F}_{s_i}\right) \bigg| \mathcal{F}_s\right).
\end{aligned}
$$

Using Lemma 7.2.1 once again

$$
\mathbf{E}\left[(Z_{i+1} - \mathbf{E}(Z_{i+1}|\mathcal{F}_{s_i}))^2 \big| \mathcal{F}_{s_i}\right] = V(Y_{s_i}),
$$

where the function V is defined by

$$
V(y) = \mathrm{Var}\left(\sum_{j=1}^{N_{s_{i+1}} - N_{s_i}} \Phi(y, U_{N_{s_i}+j})\right)
$$

and, from Exercise 40,

$$
V(y) = \lambda(s_{i+1} - s_i) \int d\nu(z) \Phi^2(y, z).
$$

Therefore

$$
\mathbf{E}\left((Z^\rho - \bar{Z}^\rho)^2 \big| \mathcal{F}_s\right) = \mathbf{E}\left(\sum_{i=0}^{m-1} \lambda(s_{i+1} - s_i) \int d\nu(z) \Phi^2(Y_{s_i}, z) \bigg| \mathcal{F}_s\right),
$$

and so, when the mesh of the partition ρ tends to 0,

$$
\mathbf{E}\left[(M_t - M_s)^2 \big| \mathcal{F}_s\right] = \mathbf{E}\left[\lambda \int_s^t du \int d\nu(z) \Phi^2(Y_u, z) \bigg| \mathcal{F}_s\right]. \tag{7.3}
$$

Since $(M_t)_{t\geq 0}$ is a square-integrable martingale, we obtain

$$
\mathbf{E}\left[(M_t - M_s)^2 \big| \mathcal{F}_s\right] = \mathbf{E}\left(M_t^2 + M_s^2 - 2M_t M_s | \mathcal{F}_s\right) = \mathbf{E}\left(M_t^2 - M_s^2 | \mathcal{F}_s\right)
$$

and equality (7.3) implies that $M_t^2 - \lambda \int_0^t du \int d\nu(z) \Phi^2(Y_u, z)$ is a martingale.

If we do not assume that Φ is bounded, but instead

$$\mathbf{E} \left(\int_0^t ds \int d\nu(z) \Phi^2(Y_s, z) \right) < +\infty,$$

for any t, we can introduce the (bounded) functions Φ^n's defined by $\Phi^n(y, z) = \inf(n, \sup(-n, \Phi(y, z)))$, and the martingales $(M_t^n)_{t \geq 0}$ defined by

$$M_t^n = \sum_{j=1}^{N_t} \Phi^n(Y_{\tau_j}, U_j) - \lambda \int_0^t ds \int \nu(dz) \Phi^n(Y_s, z).$$

It is easily seen that $\mathbf{E} \left(\int_0^t ds \int \nu(dz) \left(\Phi^n(Y_s, z) - \Phi(Y_s, z) \right)^2 \right)$ tends to 0 as n tends to infinity. It follows that the sequence $(M_t^n)_{n \geq 1}$ is Cauchy in L^2 and as M_t^n tends to M_t a.s., M_t is square-integrable and taking the limit, the lemma is satisfied for Φ. □

Lemma 7.2.3 *We keep the hypothesis and notations of Lemma 7.2.2. Let $(A_t)_{t \geq 0}$ be an adapted process such that $\mathbf{E} \left(\int_0^t A_s^2 ds \right) < +\infty$ for any t. We set $L_t = \int_0^t A_s dW_s$ and, as in Lemma 7.2.2,*

$$M_t = \sum_{j=1}^{N_t} \Phi(Y_{\tau_j}, U_j) - \lambda \int_0^t ds \int \nu(dz) \Phi(Y_s, z).$$

Then the product $L_t M_t$ is a martingale.

Proof. It is sufficient to prove the lemma for Φ bounded (the general case is proved by approximating Φ by some $\Phi^n = \inf(n, \sup(-n, \Phi))$, as in the proof of Lemma 7.2.2). Let us fix $s < t$ and denote by $\rho = (s_0 = s < s_1 < \cdots < s_m = t)$ a partition of the interval $[s, t]$. We have

$$\mathbf{E} \left[(L_t M_t - L_s M_s) | \mathcal{F}_s \right] = \mathbf{E} \left[\sum_{i=0}^{m-1} \mathbf{E} \left((L_{s_{i+1}} M_{s_{i+1}} - L_{s_i} M_{s_i}) | \mathcal{F}_{s_i} \right) \Big| \mathcal{F}_s \right].$$

On the other hand, since $(L_t)_{t \geq 0}$ and $(M_t)_{t \geq 0}$ are martingales

$$\mathbf{E} \left((L_{s_{i+1}} M_{s_{i+1}} - L_{s_i} M_{s_i}) | \mathcal{F}_{s_i} \right) = \mathbf{E} \left((L_{s_{i+1}} - L_{s_i})(M_{s_{i+1}} - M_{s_i}) | \mathcal{F}_{s_i} \right).$$

Whence

$$\mathbf{E} \left((L_t M_t - L_s M_s) | \mathcal{F}_s \right) = \mathbf{E}(\Lambda^\rho | \mathcal{F}_s)$$

with

$$\Lambda^\rho = \sum_{i=0}^{m-1} (L_{s_{i+1}} - L_{s_i})(M_{s_{i+1}} - M_{s_i}).$$

We deduce

$$|\Lambda^\rho| \leq \left(\sup_{0 \leq i \leq m-1} |L_{s_{i+1}} - L_{s_i}|\right) \sum_{i=0}^{m-1} |M_{s_{i+1}} - M_{s_i}|$$

$$\leq \sup_{0 \leq i \leq m-1} |L_{s_{i+1}} - L_{s_i}|$$

$$\left(\sum_{j=N_s+1}^{N_t} |\Phi(Y_{\tau_j}, U_j)| + \lambda \int_s^t du \int d\nu(z)|\Phi(Y_s, z)|\right)$$

$$\leq \left(\sup_{0 \leq i \leq m-1}\right)|L_{s_{i+1}} - L_{s_i}| \left(C(N_t - N_s) + \lambda C(t - s)\right),$$

with $C = \sup_{y,z} |\Phi(y, z)|$. From the continuity of $t \mapsto L_t$, we see that Λ^ρ tends almost surely to 0 as the mesh of the partition ρ tends to 0. Moreover

$$|\Lambda^\rho| \leq 2 \sup_{s \leq u \leq t} |L_u| \left(C(N_t - N_s) + \lambda C(t - s)\right).$$

The random variable $\sup_{s \leq u \leq t} |L_u|$ is in L^2 (from the Doob inequality, cf. Chapter 3, Proposition 3.3.7), as well as $N_t - N_s$. We deduce that Λ^ρ tends to 0 in L^1, and consequently

$$\mathbf{E}\left[(L_t M_t - L_s M_s)|\mathcal{F}_s\right] = 0.$$

$$\square$$

7.3 Pricing and hedging options

7.3.1 Admissible strategies

Let us go back to the model introduced at the beginning of the previous section, assuming that the U_i's are square-integrable and that

$$\mu = r - \lambda \mathbf{E}(U_1) = r - \lambda \int z\nu(dz), \tag{7.4}$$

which implies that the process $\left(\tilde{X}_t\right)_{t \geq 0} = (e^{-rt} X_t)_{t \geq 0}$ is a martingale. Notice that

$$\mathbf{E}\left(X_t^2\right) = X_0^2 \mathbf{E}\left(\exp\left(\left(\mu - \frac{\sigma^2}{2}\right)t + \sigma W_t\right) \prod_{j=1}^{N_t}(1 + U_j)\right)^2$$

and consequently, from Exercise 39,

$$\mathbf{E}\left(X_t^2\right) = X_0^2 \exp\left((\sigma^2 + 2r)t\right) \exp\left(\lambda t \mathbf{E}(U_1^2)\right).$$

Therefore the process $\left(\tilde{X}_t\right)_{t \geq 0}$ is a square-integrable martingale.

In the following, we fix a finite horizon T. A trading strategy will be defined, as in the Black-Scholes model, by an adapted process $\phi = ((H_t^0, H_t))_{0 \leq t \leq T}$, taking values in $\mathbb{R}^2$, representing the amounts of assets held over time; but, to take the jumps into account, we will constrain the processes (H_t^0) and (H_t) to be *left-continuous*. Since the process (X_t) is itself right-continuous, this means, intuitively, that one can react to the jumps only after their occurrence. This condition is the counterpart of the condition of *predictability* which is found in the discrete models (cf. Chapter 1) and which is slightly more prickly to define in continuous time.

The value at time t of the strategy ϕ is given by $V_t = H_t^0 e^{rt} + H_t X_t$ and the strategy is said to be self-financing if

$$dV_t = H_t^0 r e^{rt} dt + H_t dX_t,$$

i.e., taking into account equation (7.1), $dV_t = H_t^0 r e^{rt} dt + H_t X_t (\mu dt + \sigma dW_t)$ between the jump times and at a jump time τ_j, V_t jumps by an amount $\Delta V_{\tau_j} = H_{\tau_j} \Delta X_{\tau_j} = H_{\tau_j} U_j X_{\tau_j^-}$. Precisely, the condition of self-financing can be written as

$$V_t = H_t^0 e^{rt} + H_t X_t = V_0 + \int_0^t H_s^0 r e^{rs} ds + \int_0^t H_s X_s (\mu ds + \sigma dW_s)$$
$$+ \sum_{j=1}^{N_t} H_{\tau_j} U_j X_{\tau_j^-}. \tag{7.5}$$

For this equation to make sense, it suffices to impose the condition $\int_0^T |H_s^0| ds + \int_0^T H_s^2 ds < \infty$, a.s. (it is easily seen that $s \mapsto X_s$ is almost surely bounded). Actually, for a specific reason to be discussed later, we will impose a stronger condition of integrability on the process $(H_t)_{0 \leq t \leq T}$, by restricting the class of admissible strategies as follows:

Definition 7.3.1 *An admissible strategy is defined by a process*

$$\phi = ((H_t^0, H_t))_{0 \leq t \leq T}$$

adapted, left-continuous, with values in $\mathbb{R}^2$, satisfying equality (7.5) a.s. for all $t \in [0, T]$ and such that
$\int_0^T |H_s^0| ds < +\infty$ $\mathbf{P}$ *a.s. and* $\mathbf{E} \left(\int_0^T H_s^2 X_s^2 ds \right) < +\infty$.

Note that we do not impose any condition of non-negativity on the value of admissible strategies. The following proposition is the counterpart of Proposition 4.1.2 of Chapter 4.

Proposition 7.3.2 *Let $(H_t)_{0 \leq t \leq T}$ be an adapted, left-continuous process such that*

$$\mathbf{E} \left(\int_0^T H_s^2 X_s^2 ds \right) < \infty,$$

and let $V_0 \in \mathbb{R}$. There exists a unique process $(H_t^0)_{0 \leq t \leq T}$ such that the pair

$((H_t^0, H_t))_{0 \le t \le T}$ *defines an admissible strategy with initial value V_0. The discounted value at time t of this strategy is given by*

$$\tilde{V}_t = V_0 + \int_0^t H_s \tilde{X}_s \sigma dW_s + \sum_{j=1}^{N_t} H_{\tau_j} U_j \tilde{X}_{\tau_j^-} - \lambda \int_0^t ds H_s \tilde{X}_s \int \nu(dz) z.$$

Proof. If the pair $(H_t^0, H_t)_{0 \le t \le T}$ defines an admissible strategy, its value at time t is given by $V_t = Y_t + Z_t$, with $Y_t = V_0 + \int_0^t H_s^0 r e^{rs} ds + \int_0^t H_s X_s (\mu ds + \sigma dW_s)$ and $Z_t = \sum_{j=1}^{N_t} H_{\tau_j} U_j X_{\tau_j^-}$. Differentiating the product $e^{-rt} Y_t$,

$$e^{-rt} V_t = V_0 + \int_0^t (-r e^{-rs}) Y_s ds + \int_0^t e^{-rs} dY_s + e^{-rt} Z_t. \qquad (7.6)$$

Moreover, the product $e^{-rt} Z_t$ can be written as follows:

$$
\begin{aligned}
e^{-rt} Z_t &= \sum_{j=1}^{N_t} e^{-rt} H_{\tau_j} U_j X_{\tau_j^-} \\
&= \sum_{j=1}^{N_t} \left(e^{-r\tau_j} + \int_{\tau_j}^t (-r e^{-rs}) ds \right) H_{\tau_j} U_j X_{\tau_j^-} \\
&= \sum_{j=1}^{N_t} e^{-r\tau_j} H_{\tau_j} U_j X_{\tau_j^-} \\
&\quad + \sum_{j=1}^{N_t} \int_0^t ds 1_{\{\tau_j \le s\}} (-r e^{-rs}) H_{\tau_j} U_j X_{\tau_j^-} \\
&= \sum_{j=1}^{N_t} e^{-r\tau_j} H_{\tau_j} U_j X_{\tau_j^-} + \int_0^t ds (-r e^{-rs}) \sum_{j=1}^{N_s} H_{\tau_j} U_j X_{\tau_j^-} \\
&= \sum_{j=1}^{N_t} e^{-r\tau_j} H_{\tau_j} U_j X_{\tau_j^-} + \int_0^t (-r e^{-rs}) Z_s ds.
\end{aligned}
$$

Writing this in (7.6) and expressing dY_s, we obtain

$$
\begin{aligned}
\tilde{V}_t &= V_0 + \int_0^t (-r e^{-rs}) V_s ds + \int_0^t H_s^0 r ds + \int_0^t H_s \tilde{X}_s (\mu ds + \sigma dW_s) \\
&\quad + \sum_{j=1}^{N_t} H_{\tau_j} U_j \tilde{X}_{\tau_j^-} \\
&= V_0 - \int_0^t r \left(H_s^0 + H_s \tilde{X}_s \right) ds + \int_0^t H_s^0 r ds + \int_0^t H_s \tilde{X}_s (\mu ds + \sigma dW_s) \\
&\quad + \sum_{j=1}^{N_t} H_{\tau_j} U_j \tilde{X}_{\tau_j^-}
\end{aligned}
$$

$$\begin{aligned} = \quad & V_0 + \int_0^t H_s \tilde{X}_s ((\mu - r)ds + \sigma dW_s) \\ & + \sum_{j=1}^{N_t} H_{\tau_j} U_j \tilde{X}_{\tau_j^-}, \end{aligned}$$

which, taking into account equality (7.4), yields

$$\tilde{V}_t = V_0 + \int_0^t H_s \tilde{X}_s \sigma dW_s + \sum_{j=1}^{N_t} H_{\tau_j} U_j \tilde{X}_{\tau_j^-} - \lambda \int_0^t ds H_s \tilde{X}_s \int \nu(dz)z.$$

It is clear then that if V_0 and (H_t) are given, the unique process (H_t^0) such that $((H_t^0, H_t))_{0 \le t \le T}$ is an admissible strategy with initial value V_0 is given by

$$\begin{aligned} H_t^0 \quad = \quad & \tilde{V}_t - H_t \tilde{X}_t \\ = \quad & -H_t \tilde{X}_t + V_0 + \int_0^t H_s \tilde{X}_s \sigma dW_s + \sum_{j=1}^{N_t} H_{\tau_j} U_j \tilde{X}_{\tau_j^-} \\ & -\lambda \int_0^t ds H_s \tilde{X}_s \int \nu(dz)z. \end{aligned}$$

From this formula, we see that the process (H_t^0) is adapted, has left-hand limit at any point and is such that $H_t^0 = H_{t-}^0$. This last property is straightforward if t is not a jump time τ_j and if t is some τ_j, we have

$$H_{\tau_j}^0 - H_{\tau_j^-}^0 = -H_{\tau_j} \Delta \tilde{X}_{\tau_j} + H_{\tau_j} U_j \tilde{X}_{\tau_j^-} = 0.$$

It is also obvious that $\int_0^T |H_t^0| dt < \infty$ almost surely. Moreover, writing $H_t^0 e^{rt} + H_t X_t = e^{rt} \left(H_t^0 + H_t \tilde{X}_t \right)$ and integrating by parts as above we see that

$$((H_t^0, H_t))_{0 \le t \le T}$$

defines an admissible strategy with initial value V_0. $\qquad \square$

Remark 7.3.3 The condition $\mathbf{E} \left(\int_0^T H_s^2 \tilde{X}_s^2 ds \right) < \infty$ implies that the discounted value $(\tilde{V}_t)$ of an admissible strategy is a square-integrable martingale. This results from the expression in Proposition 7.3.2 and Lemma 7.2.2, applied with the continuous process with left-hand limit defined by $Y_t = (H_t, \tilde{X}_{t-})$ (note that in the integral with respect to ds, one can substitute $\tilde{X}_s$ for $\tilde{X}_{s-}$ because there is only finitely many discontinuities).

7.3.2 *Pricing*

Let us consider a European option with maturity T, defined by a random variable h, $\mathcal{F}_T$-measurable and square-integrable. To clarify, let us stand from the writer's point of view. He sells the option at a price V_0 at time 0 and then follows an admissible strategy between times 0 and T. From Proposition 7.3.2, this strategy is completely determined by the process $(H_t)_{0 \le t \le T}$ representing the amount of the risky asset. If V_t represents the value of this strategy at time t, the hedging mismatch at maturity is given by $h - V_T$. If this quantity is non-negative, the writer of the option loses money, otherwise he earns some. A way of measuring the risk is to introduce the quantity

$$R_0^T = \mathbf{E}\left(\left(e^{-rT}(h - V_T) \right)^2 \right).$$

Since, from Remark 7.3.3, the discounted value $(\tilde{V}_t)$ is a martingale, we have $\mathbf{E}\left(e^{-rT}V_T \right) = V_0$. Applying the identity $\mathbf{E}(Z^2) = (\mathbf{E}(Z))^2 + \mathbf{E}\left([Z - \mathbf{E}(Z)]^2 \right)$ to the random variable $Z = e^{-rT}(h - V_T)$, we obtain

$$R_0^T = \left(\mathbf{E}(e^{-rT}h) - V_0 \right)^2 + \mathbf{E}\left(e^{-rT}h - \mathbf{E}(e^{-rT}h) - (\tilde{V}_T - V_0) \right)^2. \quad (7.7)$$

Proposition 7.3.2 shows that the quantity $\tilde{V}_T - V_0$ depends only on (H_t) (and not on V_0). If the writer of the option tries to minimise the risk R_0^T, he will ask for a premium $V_0 = \mathbf{E}(e^{-rT}h)$. So it appears that $\mathbf{E}(e^{-rT}h)$ is the initial value of any strategy designed to minimise the risk at maturity and this is what we will take as a definition of the price of the option associated with h. By a similar argument, we see that an agent selling the option at time $t > 0$, who wants to minimise the quantity $R_t^T = \mathbf{E}\left(\left(e^{-r(T-t)}(h - V_T) \right)^2 | \mathcal{F}_t \right)$, will ask for a premium $V_t = \mathbf{E}\left(e^{-r(T-t)}h | \mathcal{F}_t \right)$. We will take this quantity to define the price of the option at time t.

7.3.3 Prices of calls and puts

Before tackling the problem of hedging, we try to give an explicit expression for the price of the call or the put with strike price K. We will assume therefore that h can be written as $f(X_T)$, with $f(x) = (x - K)_+$ or $f(x) = (K - x)_+$. As we saw earlier, the price of the option at time t is given by

$$\mathbf{E}\left(e^{-r(T-t)}f(X_T) | \mathcal{F}_t \right)$$

$$= \mathbf{E}\left(e^{-r(T-t)}f\left(X_t e^{(\mu - \sigma^2/2)(T-t) + \sigma(W_T - W_t)} \prod_{j=N_t+1}^{N_T} (1 + U_j) \right) \Big| \mathcal{F}_t \right)$$

$$= \mathbf{E}\left(e^{-r(T-t)}f\left(X_t e^{(\mu - \sigma^2/2)(T-t) + \sigma(W_T - W_t)} \prod_{j=1}^{N_T - N_t} (1 + U_{N_t+j}) \right) \Big| \mathcal{F}_t \right).$$

From Lemma 7.2.1 and this equality, we deduce that

$$\mathbf{E}\left(e^{-r(T-t)}f(X_T) | \mathcal{F}_t \right) = F(t, X_t),$$

with

$$F(t,x) = \mathbf{E}\left(e^{-r(T-t)}f\left(xe^{(\mu-\sigma^2/2)(T-t)+\sigma W_{T-t}}\prod_{j=1}^{N_{T-t}}(1+U_j)\right)\right)$$

$$= \mathbf{E}\left(e^{-r(T-t)}f\left(xe^{(r-\lambda\mathbf{E}(U_1)-\sigma^2/2)(T-t)+\sigma W_{T-t}}\prod_{j=1}^{N_{T-t}}(1+U_j)\right)\right).$$

Note that if we introduce the function

$$F_0(t,x) = \mathbf{E}\left(e^{-r(T-t)}f\left(xe^{(r-\sigma^2/2)(T-t)+\sigma W_{T-t}}\right)\right),$$

which gives the price of the option for the Black-Scholes model, we have

$$F(t,x) = \mathbf{E}\left(F_0\left(t, xe^{-\lambda(T-t)\mathbf{E}(U_1)}\prod_{j=1}^{N_{T-t}}(1+U_j)\right)\right). \tag{7.8}$$

Since N_{T-t} is a random variable independent of the U_j's, following a Poisson law with parameter $\lambda(T-t)$, we can also write

$$F(t,x) = \sum_{n=0}^{\infty}\mathbf{E}\left(F_0\left(t, xe^{-\lambda(T-t)\mathbf{E}(U_1)}\prod_{j=1}^{n}(1+U_j)\right)\right)\frac{e^{-\lambda(T-t)}\lambda^n(T-t)^n}{n!}.$$

Each term of this series can be computed numerically if we know how to simulate the law of the U_j's. For some laws, the mathematical expectation in the formula can be calculated explicitly (cf. Exercise 42).

7.3.4 Hedging of calls and puts

Let us examine the hedging problem for an option $h = f(X_T)$, with $f(x) = (x - K)_+$ or $f(x) = (K - x)_+$. We have seen that the initial value of any admissible strategy aiming at minimising the risk R_0^T at maturity is given by $V_0 = \mathbf{E}(e^{-rT}h) = F(0, X_0)$. For such a strategy, equality (7.7) yields

$$R_0^T = \mathbf{E}\left(e^{-rT}h - \tilde{V}_T\right)^2.$$

Now we determine a process $(H_t)_{0\leq t\leq T}$ for the quantities of the risky asset to be held in portfolio to minimise R_0^T. To do so, we need the following proposition.

Proposition 7.3.4 *Let V_t be the value at time t of an admissible strategy with initial value $V_0 = \mathbf{E}\left(e^{-rT}f(X_T)\right) = F(0, X_0)$, determined by a process $(H_t)_{0\leq t\leq T}$ for the quantities of the risky asset. The quadratic risk at maturity*

$R_0^T = \mathbf{E}\left(e^{-rT}(f(X_T) - V_T)\right)^2$ *is given by the following formula:*

$$R_0^T = \mathbf{E}\left(\int_0^T \left(\frac{\partial F}{\partial x}(s, X_s) - H_s\right)^2 \tilde{X}_s^2 \sigma^2 ds\right.$$

$$\left. + \int_0^T \lambda \int \nu(dz) e^{-2rs} \left(F(s, X_s(1+z)) - F(s, X_s) - H_s z X_s\right)^2 ds\right).$$

Proof. From Proposition 7.3.2, we have, for $t \leq T$,

$$\tilde{V}_t = F(0, X_0) + \int_0^t \sigma H_s \tilde{X}_s dW_s + \sum_{j=1}^{N_t} H_{\tau_j} U_j \tilde{X}_{\tau_j^-} - \lambda \int_0^t ds \tilde{X}_s H_s \mathbf{E}(U_1).$$

$$(7.9)$$

On the other hand, we have $\tilde{h} = e^{-rT} f(X_T) = e^{-rT} F(T, X_T)$. Let us introduce the function $\tilde{F}$ defined by

$$\tilde{F}(t, x) = e^{-rt} F(t, xe^{rt}),$$

so that $\tilde{F}(t, \tilde{X}_t) = \mathbf{E}\left(\tilde{h}|\mathcal{F}_t\right)$. It emerges that $\tilde{F}(t, \tilde{X}_t)$ is the discounted price of the option at time t. We deduce easily (exercise) from formula (7.8) that $\tilde{F}(t, x)$ is C^2 on $[0, T[\times\mathbb{R}^+$ and, writing down the Itô formula between the jump times, we obtain

$$\tilde{F}(t, \tilde{X}_t) =$$

$$F(0, X_0) + \int_0^t \frac{\partial \tilde{F}}{\partial s}(s, \tilde{X}_s) ds + \int_0^t \frac{\partial \tilde{F}}{\partial x}(s, \tilde{X}_s) \tilde{X}_s(-\lambda \mathbf{E}(U_1) ds + \sigma dW_s)$$

$$+ \frac{1}{2}\int_0^t \frac{\partial^2 \tilde{F}}{\partial x^2}(s, \tilde{X}_s)\sigma^2 \tilde{X}_s^2 ds + \sum_{j=1}^{N_t} \tilde{F}(\tau_j, \tilde{X}_{\tau_j}) - \tilde{F}(\tau_j, \tilde{X}_{\tau_j^-}). \tag{7.10}$$

Remark that the function $\tilde{F}(t, x)$ is Lipschitz of order 1 with respect to x, since

$$|F(t, x) - F(t, y)|$$

$$\leq \mathbf{E}\left(e^{-r(T-t)} \left| f\left(xe^{(r-\lambda\mathbf{E}(U_1)-\sigma^2/2)(T-t)+\sigma W_{T-t}} \prod_{j=1}^{N_{T-t}}(1+U_j)\right)\right.\right.$$

$$\left.\left. - f\left(ye^{(r-\lambda\mathbf{E}(U_1)-\sigma^2/2)(T-t)+\sigma W_{T-t}} \prod_{j=1}^{N_{T-t}}(1+U_j)\right)\right|\right)$$

$$\leq |x - y|\mathbf{E}\left(e^{-\lambda\mathbf{E}(U_1)(T-t)} e^{\sigma W_{T-t}-(\sigma^2/2)(T-t)} \prod_{j=1}^{N_{T-t}}(1+U_j)\right)$$

$$= |x - y|.$$

It follows that

$$\mathbf{E}\left(\int_0^t ds \int \nu(dz)\left(\tilde{F}(s,\tilde{X}_s(1+z)) - \tilde{F}(s,\tilde{X}_s)\right)^2\right)$$
$$\leq \mathbf{E}\left(\int_0^t ds\,\tilde{X}_s^2\int \nu(dz)z^2\right)$$
$$< +\infty,$$

which, from Lemma 7.2.2, implies that the process

$$M_t = \sum_{j=1}^{N_t}\tilde{F}(\tau_j,\tilde{X}_{\tau_j}) - \tilde{F}(\tau_j,\tilde{X}_{\tau_j}-)$$
$$- \lambda\int_0^t ds\int\left(\tilde{F}(s,\tilde{X}_s(1+z)) - \tilde{F}(s,\tilde{X}_s)\right)d\nu(z)$$

is a square-integrable martingale. We also know that $\tilde{F}(t,\tilde{X}_t)$ is a martingale. Therefore the process $\tilde{F}(t,\tilde{X}_t) - M_t$ is also a martingale and, from equality (7.10), it is an Itô process. From Exercise 16 of Chapter 3, it can be written as a stochastic integral. Whence

$$\tilde{F}(t,\tilde{X}_t) - M_t = F(0,X_0) + \int_0^t \frac{\partial\tilde{F}}{\partial x}(s,\tilde{X}_s)\tilde{X}_s\sigma dW_s. \qquad (7.11)$$

Gathering equalities (7.9) and (7.11), we get

$$\tilde{h} - \tilde{V}_T = M_T^{(1)} + M_T^{(2)},$$

with

$$M_t^{(1)} = \int_0^t\left(\frac{\partial\tilde{F}}{\partial x}(s,\tilde{X}_s) - H_s\right)\sigma\tilde{X}_s dW_s$$

and

$$M_t^{(2)} = \sum_{j=1}^{N_t}\left(\tilde{F}(\tau_j,\tilde{X}_{\tau_j}) - \tilde{F}(\tau_j,\tilde{X}_{\tau_j}-) - H_{\tau_j}U_j\tilde{X}_{\tau_j}-\right)$$
$$-\lambda\int_0^t ds\int d\nu(z)\left(\tilde{F}(s,\tilde{X}_s(1+z)) - \tilde{F}(s,\tilde{X}_s) - H_s z\tilde{X}_s\right).$$

From Lemma 7.2.3, $M_t^{(1)}M_t^{(2)}$ is a martingale and consequently

$$\mathbf{E}\left(M_t^{(1)}M_t^{(2)}\right) = M_0^{(1)}M_0^{(2)} = 0.$$

Whence

$$\mathbf{E}\left(\tilde{h} - \tilde{V}_T\right) = \mathbf{E}((M_T^{(1)})^2) + \mathbf{E}((M_T^{(2)})^2)$$
$$= \mathbf{E}\left(\int_0^T\left(\frac{\partial\tilde{F}}{\partial x}(s,\tilde{X}_s) - H_s\right)^2\tilde{X}_s^2\sigma^2 ds\right) + \mathbf{E}((M_T^{(2)})^2),$$

and applying Lemma 7.2.2 again

$$\mathbf{E}((M_T^{(2)})^2)$$
$$= \mathbf{E}\left(\lambda \int_0^T ds \int \nu(dz)\left(\tilde{F}(s,\tilde{X}_s(1+z)) - \tilde{F}(s,\tilde{X}_s) - H_s z \tilde{X}_s\right)^2\right).$$

The risk at maturity is then given by

$$R_0^T = \mathbf{E}\left(\int_0^T \left(\frac{\partial \tilde{F}}{\partial x}(s,\tilde{X}_s) - H_s\right)^2 \tilde{X}_s^2 \sigma^2 ds\right.$$
$$\left. + \int_0^T \lambda \int \nu(dz)\left(\tilde{F}(s,\tilde{X}_s(1+z)) - \tilde{F}(s,\tilde{X}_s) - H_s z \tilde{X}_s\right)^2 ds\right).$$

$$\square$$

It follows that the minimal risk is obtained when H_s satisfies **P** a.s.

$$\left(\frac{\partial \tilde{F}}{\partial x}(s,\tilde{X}_s) - H_s\right)\tilde{X}_s^2 \sigma^2$$
$$+ \ \lambda \int \nu(dz)\left(\tilde{F}(s,\tilde{X}_s(1+z)) - \tilde{F}(s,\tilde{X}_s) - H_s z \tilde{X}_s\right)z\tilde{X}_s = 0.$$

It suffices indeed to minimise the integrand with respect to ds. It yields, since $(H_t)_{t\geq 0}$ must be left-continuous,

$$H_s = \Delta(s, X_{s^-}),$$

with

$$\Delta(s,x) = \frac{1}{\sigma^2 + \lambda \int \nu(dz)z^2}\left(\sigma^2 \frac{\partial F}{\partial x}(s,x)\right.$$
$$\left. +\lambda \int \nu(dz)z\frac{(F(s,x(1+z)) - F(s,x))}{x}\right).$$

In this way, we obtain a process which satisfies $\mathbf{E}\left(\int_0^T H_s^2 \tilde{X}_s^2 ds\right) < +\infty$ and which determines therefore an admissible strategy minimising the risk at maturity. Note that if there is no jump ($\lambda = 0$), we recover the hedging formula for the Black-Scholes model and, in this case, we know that the hedging is perfect, i.e. $R_0^T = 0$. But, when there are jumps, the minimal risk is generally positive (cf. Exercise 43 and Chateau (1990)).

Remark 7.3.5 The formulae we obtain indicate that calculations are still possible for models with jump. It remains to identify parameters and the law of the U_i's. As for the volatility in the Black-Scholes model, we can distinguish two approaches: (1) a statistical approach, from the historical data and (2) an implied approach, from the market data, in other words from the prices of options quoted on an organised market. In the second approach, the models with jump, which involve several parameters, give a better 'fit' to the market prices.

Notes: The financial models with jumps were introduced by Merton (1976). The approach used in this chapter is based on Föllmer and Sondermann (1986), CERMA (1988) and Bouleau and Lamberton (1989). The approach we have chosen relies heavily on the assumption that the discounted stock price is a martingale. This assumption is rather arbitrary. Moreover, the use of variance as a measure of risk is questionable. Therefore, the reader is urged to consult the recent literature dealing with incomplete markets, especially Föllmer and Schweizer (1991), Schweizer (1992, 1993, 1994), El Karoui and Quenez (1995).

7.4 Exercises

Exercise 39 Let $(V_n)_{n\geq 1}$ be a sequence of non-negative, independent and identically distributed random variables and let N be a random variable with values in $\mathbf{N}$, following a Poisson law with parameter λ, independent of the sequence $(V_n)_{n\geq 1}$. Show that

$$\mathbf{E}\left(\prod_{n=1}^{N} V_n\right) = e^{\lambda(\mathbf{E}(V_1)-1)}.$$

Exercise 40 Let $(V_n)_{n\geq 1}$ be a sequence of independent, identically distributed, integrable random variables and let N be a random variable taking values in $\mathbf{N}$, integrable and independent of the sequence (V_n). We set $S = \sum_{n=1}^{N} V_n$ (with the convention $\sum_{n=1}^{0} = 0$).

1. Prove that S is integrable and that $\mathbf{E}(S) = \mathbf{E}(N)\mathbf{E}(V_1)$.

2. We assume N and V_1 to be square-integrable. Then show that S is square-integrable and that its variance is $\mathrm{Var}(S) = \mathbf{E}(N)\mathrm{Var}(V_1) + \mathrm{Var}(N)\left(\mathbf{E}(V_1)\right)^2$.

3. Deduce that if N follows a Poisson law with parameter λ, $\mathbf{E}(S) = \lambda\mathbf{E}(V_1)$ and $\mathrm{Var}(S) = \lambda\mathbf{E}\left(V_1^2\right)$.

Exercise 41 The hypothesis and notations are those of Exercise 40. We suppose that the V_j's take values in $\{\alpha, \beta\}$, with $\alpha, \beta \in \mathbb{R}$ and we set $p = \mathbf{P}(V_1 = \alpha) = 1 - \mathbf{P}(V_1 = \beta)$. Prove that S has the same law as $\alpha N_1 + \beta N_2$, where N_1 and N_2 are two independent random variables following a Poisson law with respective parameters λp and $(1 - p)\lambda$.

Exercise 42

1. We suppose, with the notations of Section 7.3, that U_1 takes values in $\{a, b\}$, with $p = \mathbf{P}(U_1 = a) = 1 - \mathbf{P}(U_1 = b)$. Write the price formula (7.8) as a double series where each term is calculated from the Black-Scholes formulae (hint: use Exercise 41).

2. Now we suppose that U_1 has the same law as $e^g - 1$, where g is a normal variable with mean m and variance σ^2. Write the price formula (7.8) as a series of terms calculated from the Black-Scholes formulae (for some interest rates and volatilities to be given).

Exercise 43 The objective of this exercise is to show that there is no perfect hedging of calls and puts for the models with jumps we studied in this chapter. We consider a model in which $\sigma > 0$, $\lambda > 0$ and $\mathbf{P}\,(U_1 \neq 0) > 0$.

1. From Proposition 7.3.4, show that if there is a perfect hedging scheme then, for ds almost every s and for ν almost every z, we have

$$\mathbf{P} \text{ a.s.} \quad zX_s \frac{\partial F}{\partial x}(s, X_s) = F\,(s, X_s(1 + z)) - F(s, X_s).$$

2. Show that the law of X_s has (for $s > 0$) a positive density on $]0, \infty[$. It may be worth noticing that if Y has a density g and if Z is a random variable independent of Y with values in $]0, \infty[$, the random variable YZ has the density $\int d\mu(z)(1/z)g(y/z)$, where μ is the law of Z.

3. Under the same assumptions as in the first question, show that there exists $z \neq 0$ such that for $s \in [0, T[$ and $x \in]0, \infty[$,

$$\frac{\partial F}{\partial x}(s, x) = \frac{F\,(s, x(1 + z)) - F(s, x)}{zx}.$$

   Deduce (using the convexity of F with respect to x) that, for $s \in [0, T]$, the function $x \mapsto F(s, x)$ is linear.

4. Conclude. It may be noticed that, in the case of the put, the function $x \mapsto F(s, x)$ is non-negative and decreasing on $]0, \infty[$.

8

Simulation and algorithms
for financial models

8.1 Simulation and financial models

In this chapter, we describe some methods which can be used to simulate financial models and compute prices. When we can write the option price as the expectation of a random variable that can be simulated, Monte Carlo methods can be used. Unfortunately these methods are inefficient and are only used if there is no closed-form solution for the price of the option. Simulations are also useful to evaluate complex hedging strategies (example: find the impact of hedging a portfolio every ten days instead of every day, see Exercise 46).

8.1.1 The Monte Carlo method

The problem of simulation can be presented as follows. We consider a random variable with law $\mu(dx)$ and we would like to generate a sequence of independent trials, $X_1, \ldots, X_n, \ldots$ with common distribution μ. Applying the law of large numbers, we can assert that if f is a μ-integrable function

$$\lim_{N \to +\infty} \frac{1}{N} \sum_{1 \leq n \leq N} f(X_n) = \int f(x)\mu(dx). \tag{8.1}$$

To implement this method on a computer, we proceed as follows. We suppose that we know how to build a sequence of numbers $(U_n)_{n \geq 1}$ which is the realisation of a sequence of independent, uniform random variables on the interval $[0, 1]$ and we look for a function $(u_1, \ldots, u_p) \mapsto F(u_1, \ldots, u_p)$ such that the random variable $F(U_1, \ldots, U_p)$ has the desired law $\mu(dx)$. The sequence of random variables $(X_n)_{n \geq 1}$ where $X_n = F(U_{(n-1)p+1}, \ldots, U_{np})$ is then a sequence of independent random variables following the required law μ. For example, we can apply (8.1) to the functions $f(x) = x$ and $f(x) = x^2$ to estimate the first and second-order moments of X (provided $\mathbf{E}(|X|^2)$ is finite).

The sequence $(U_n)_{n \geq 1}$ is obtained in practice from successive calls to a pseudo-random number generator. Most languages available on modern computers provide a random function, already coded, which returns either a pseudo-random number between 0 and 1, or a random integer in a fixed interval (this function is called `rand()` in C ANSI, `random` in Turbo Pascal).

Remark 8.1.1 The function F can depend in some cases (in particular when it comes to simulate stopping times), on the whole sequence $(U_n)_{n \geq 1}$, and not only on a fixed number of U_i's. The previous method can still be used if we can simulate X from an almost surely finite number of U_i's, this number being possibly random. This is the case, for example, for the simulation of a Poisson random variable (see page 163).

8.1.2 Simulation of a uniform law on $[0, 1]$

We explain how to build random number generators because very often, those available with a certain compiler are not entirely satisfactory.

The simplest and most common method is to use the linear congruential generator. A sequence $(x_n)_{n \geq 0}$ of integers between 0 and $m - 1$ is generated as follows:
$$\begin{cases} x_0 = \text{initial value} \in \{0, 1, \ldots, m - 1\} \\ x_{n+1} = ax_n + b \ (\text{modulo } m), \end{cases}$$
a, b, m being integers to be chosen cautiously in order to obtain satisfactory characteristics for the sequence. Sedgewick (1987) advocates the following choice:
$$\begin{cases} a &=& 31415821 \\ b &=& 1 \\ m &=& 10^8. \end{cases}$$

This method enables us to simulate pseudo-random integers between 0 and $m - 1$; to obtain a random real-valued number between 0 and 1 we divide this random integer by m.

```
const
    m   = 100000000;
    m1  =     10000;
    b   =  31415821;

var a : integer;

function Mult(p, q: integer) : integer;
    (* Multiplies p by q, avoiding 'overflows' *)
    var p1, p0, q1, q0 : integer;
begin
    p1 := p div m1;p0 := p mod m1;
    q1 := q div m1;q1 := q mod m1;
    Mult := (((p0*q1 + p1*q0) mod m1)*m1 + p0*q0) mod m;
end;

function Random   : real;
begin
    a := (Mult(a, b) + 1) mod m;
    Random := a/m;
end;
```

The previous generator provides reasonable results in common cases. However, it might happen that its period (here $m = 10^8$) is not big enough. Then it is possible to create random number generators with an arbitrary long period by increasing m. The interested reader will find much information on random number generators and computer procedures in Knuth (1981) and L'Ecuyer (1990).

8.1.3 Simulation of random variables

The probability laws we have used for financial models are mainly Gaussian laws (in the case of continuous models) and exponential and Poisson laws (in the case of models with jumps). We give some methods to simulate each of these laws.

Simulation of a Gaussian law

A classical method to simulate Gaussian random variables is based on the observation (see Exercise 44) that if (U_1, U_2) are two independent uniform random variables on $[0, 1]$

$$\sqrt{-2\log(U_1)}\cos(2\pi U_2)$$

follows a standard Gaussian law (i.e. zero-mean and with variance 1).

To simulate a Gaussian random variable with mean m and variance σ, it suffices to set $X = m + \sigma g$, where g is a standard Gaussian random variable.

```
function Gaussian(m,sigma : real) : real;
begin
  gaussian := m + sigma * sqrt(-2.0 * log(Random)) * cos(2.0 * pi
    * Random);
end;
```

Simulation of an exponential law

We recall that a random variable X follows an exponential law with parameter μ if its law is

$$1_{\{x \geq 0\}}\mu e^{-\mu x}dx.$$

We can simulate X noticing that, if U follows a uniform law on $[0, 1]$, $\log(U)/\mu$ follows an exponential law with parameter μ.

```
function exponential( mu : real) : real;
begin
  exponential := - log(Random) / mu;
end;
```

Remark 8.1.2 This method of simulation of the exponential law is a particular case of the so-called 'inverse distribution function' method (for this matter, see Exercise 45).

Simulation of a Poisson random variable

A Poisson random variable is a variable with values in $\mathbf{N}$ such that

$$\mathbf{P}(X = n) = e^{-\lambda}\frac{\lambda^n}{n!}, \quad \text{if } n \geq 0.$$

We have seen in Chapter 7 that if $(T_i)_{i\geq1}$ is a sequence of exponential random variables with parameter λ, then the law of $N_t = \sum_{n\geq1} n1_{\{T_1+\cdots+T_n\leq t<T_1+\cdots+T_{n+1}\}}$ is a Poisson law with parameter λt. Thus N_1 has the same law as the variable X we want to simulate. On the other hand, it is always possible to write exponential variables T_i as $-\log(U_i)/\lambda$, where the $(U_i)_{i\geq1}$'s are independent random variables following the uniform law on $[0,1]$. N_1 can be written as

$$N_1 = \sum_{n\geq1} n1_{\{U_1U_2...U_{n+1}\leq e^{-\lambda}<U_1U_2...U_n\}}.$$

This leads to the following algorithm to simulate a Poisson random variable.

```
function Poisson(lambda : real) : integer;
var
 u : real;
 n : integer;
begin
  a := exp(-lambda);
  u := Random;
  n := 0;
  while u > a do begin
    u := u * Random;
    n := n + 1
  end;
  Poisson := n
end;
```

For the simulation of laws not mentioned above or for other methods of simulation of the previous laws, one may refer to Rubinstein (1981).

Simulation of Gaussian vectors

Multidimensional models will generally involve Gaussian processes with values in $\mathbb{R}^n$. The problem in simulating Gaussian vectors (see Section A.1.2 of the Appendix for the definition of a Gaussian vector) is then essential. We give a method of simulation for this kind of random variables.

We will suppose that we want to simulate a Gaussian vector $(X_1,\ldots,X_n)$ whose law is characterised by the vector of its means $m = (m_1,\ldots,m_n) = (\mathbf{E}(X_1),\ldots,\mathbf{E}(X_n))$ and its variance matrix $\Gamma = (\sigma_{ij})_{1\leq i\leq n,1\leq j\leq n}$ where $\sigma_{ij} = \mathbf{E}(X_iX_j) - \mathbf{E}(X_i)\mathbf{E}(X_j)$. The matrix Γ is positive definite and we will assume that it is invertible. We can find the square root of Γ, in other words a matrix A, such that $A \times {}^tA = \Gamma$. As Γ is invertible so is A, and we can consider the vector $Z = A^{-1}(X - m)$. It is easily verified that this vector is a Gaussian vector with zero-mean. Moreover, its variance matrix is given by

$$\mathbf{E}(Z_iZ_j) = \sum_{1\leq k\leq n,1\leq l\leq n} \mathbf{E}\left(A_{ik}^{-1}(X_k - m_k)A_{jl}^{-1}(X_l - m_l)\right)$$

$$= \sum_{1\leq k\leq n,1\leq l\leq n} \mathbf{E}\left(A_{ik}^{-1}A_{jl}^{-1}\sigma_{kl}\right)$$

$$= (A^{-1}\Gamma({}^tA)^{-1})_{ij} = (A^{-1}A^t A({}^tA)^{-1})_{ij} = Id.$$

Z is therefore a Gaussian vector with zero-mean and a variance matrix equal to the identity. The law of the vector Z is the law of n independent standard normal variables. The law of the vector $X = m + AZ$ can then be simulated in the following manner:

- Derive the square root of the matrix Γ, say A.

- Simulate n independent standard normal variables $G = (g_1, \ldots, g_n)$.

- Compute $m + AG$.

Remark 8.1.3 To derive the square root of Γ, we may assume that A is upper-triangular; then there is a unique solution to the equation $A \times {}^t A = \Gamma$. This method of calculation of the square root is called Cholesky's method (for a complete algorithm see Ciarlet (1988)).

8.1.4 Simulation of stochastic processes

The methods delineated previously enable us to simulate a random variable, in particular the value of a stochastic process at a given time. Sometimes we need to know how to simulate the whole path of a process (for example, when we are studying the dynamics through time of the value of a portfolio of options, see Exercise 47). This section suggests some simple tricks to simulate paths of processes.

Simulation of a Brownian motion

We distinguish two methods to simulate a Brownian motion $(W_t)_{t \geq 0}$. The first one consists in 'renormalising' a random walk. Let $(X_i)_{i \geq 0}$ be a sequence of independent, identically distributed random walks with law $\mathbf{P}(X_i = 1) = 1/2$, $\mathbf{P}(X_i = -1) = 1/2$. Then we have $\mathbf{E}(X_i) = 0$ and $\mathbf{E}(X_i^2) = 1$. We set $S_n = X_1 + \cdots + X_n$; then we can 'approximate' the Brownian motion by the process $(X_t^n)_{t \geq 0}$ where

$$X_t^n = \frac{1}{\sqrt{n}} S_{[nt]}$$

where $[x]$ is the largest integer less than or equal to x. This method of simulation of the Brownian motion is partially justified in Exercise 48.

In the second method, we notice that, if $(g_i)_{i \geq 0}$ is a sequence of independent standard normal random variables, if $\Delta t > 0$ and if we set

$$\begin{cases} S_0 = 0 \\ S_{n+1} - S_n = g_n \end{cases}$$

then the law of $(\sqrt{\Delta t} S_0, \sqrt{\Delta t} S_1, \ldots, \sqrt{\Delta t} S_n)$ is identical to the law of

$$(W_0, W_{\Delta t}, W_{2\Delta t}, \ldots, W_{n\Delta t}).$$

The Brownian motion can be approximated by $X_t^n = \sqrt{\Delta t} S_{[t/\Delta t]}$.

Simulation of stochastic differential equations

There are many methods, some of them very sophisticated, to simulate the solution
of a stochastic differential equation; the reader is referred to Pardoux and Talay
(1985) or Kloeden and Platen (1992) for a review of these methods. Here we
present only the basic method, the so-called 'Euler approximation'. The principle
is the following: consider the stochastic differential equation

$$\begin{cases} X_0 & = & x \\ dX_t & = & b(X_t)dt + \sigma(X_t)dW_t. \end{cases}$$

We discretise time by a fixed mesh Δt. Then we can construct a discrete-time
process $(S_n)_{n \geq 0}$ approximating the solution of the stochastic differential equation
at times $n\Delta t$, setting

$$\begin{cases} S_0 & = & x \\ S_{n+1} - S_n & = & \left\{ b(S_n)\Delta t + \sigma(S_n)\left(W_{(n+1)\Delta t} - W_{n\Delta t}\right) \right\}. \end{cases}$$

If $X_t^n = S_{[t/\Delta t]}$, $(X_t^n)_{t \geq 0}$ approximates $(X_t)_{t \geq 0}$ in the following sense:

Theorem 8.1.4 *For any $T > 0$*

$$\mathbf{E}\left(\sup_{t \leq T} |X_t^n - X_t|^2 \right) \leq C_T \Delta t,$$

C_T *being a constant depending only on T.*

A proof of this result (as well as other schemes of discretisation of stochastic
differential equations) can be found in Chapter 7 of Gard (1988).

The law of the sequence $\left(W_{(n+1)\Delta t} - W_{n\Delta t}\right)_{n \geq 0}$ is the law of a sequence
of independent normal random variables with zero-mean and variance Δt. In a
simulation, we substitute $g_n\sqrt{\Delta t}$ to $\left(W_{(n+1)\Delta t} - W_{n\Delta t}\right)$ where $(g_n)_{n \geq 0}$ is a
sequence of independent standard normal variables. The approximating sequence
$(S_n')_{n \geq 0}$ is in this case defined by

$$\begin{cases} S_0' & = & x \\ S_{n+1}' & = & S_n' + \Delta t\, b(S_n') + \sigma(S_n')g_n\sqrt{\Delta t}. \end{cases}$$

Remark 8.1.5 We can substitute to the sequence of independent Gaussian random
variables $(g_i)_{i \geq 0}$ a sequence of independent random variables $(U_i)_{i \geq 0}$, such that
$\mathbf{P}(U_i = 1) = \mathbf{P}(U_i = -1) = 1/2$. Nevertheless, in this case, it must be noticed
that the convergence is different from that found in Theorem 8.1.4. There is still
a theorem of convergence, but it applies to the laws of the processes. Kushner
(1977) and Pardoux and Talay (1985) can be consulted for some explanations on
this kind of convergence and many results on discretisation in law for stochastic
differential equations.

An application to the Black-Scholes model

In the case of the Black-Scholes model, we want to simulate the solution of the equation

$$\begin{cases} X_0 & = & x \\ dX_t & = & X_t(rdt + \sigma dW_t). \end{cases}$$

Two approaches are available. The first consists in using the Euler approximation. We set

$$\begin{cases} S_0 & = & x \\ S_{n+1} & = & S_n(1 + r\Delta t + \sigma g_n\sqrt{\Delta t}), \end{cases}$$

and simulate X_t by $X_t^n = S_{[t/\Delta t]}$. The other method consists in using the explicit expression of the solution

$$X_t = x \exp\left(rt - \frac{\sigma^2}{2}t + \sigma W_t\right)$$

and simulating the Brownian motion by one of the methods presented previously. In the case where we simulate the Brownian motion by $\sqrt{\Delta t}\sum_{i=1}^{n} g_i$, we obtain

$$S_n = x \exp\left((r - \sigma^2/2)n\Delta t + \sigma\sqrt{\Delta t}\sum_{i=1}^{n} g_i\right). \tag{8.2}$$

We always approximate X_t by $X_t^n = S_{[t/\Delta t]}$.

Remark 8.1.6 We can also replace the Gaussian random variables g_i by some Bernouilli variables with values $+1$ or -1 with probability $1/2$ in (8.2); we obtain a binomial-type model close to the Cox-Ross-Rubinstein model used in Section 5.3.3 of Chapter 5.

Simulation of models with jumps

We have investigated in Chapter 7 an extension of the Black-Scholes model with jumps; we describe now a method to simulate this process. We take the notations and the hypothesis of Chapter 7, Section 7.2. The process $(X_t)_{t\geq 0}$ describing the dynamics of the asset is

$$X_t = x \left(\prod_{j=1}^{N_t}(1 + U_j)\right) e^{(\mu - \sigma^2/2)t + \sigma W_t}, \tag{8.3}$$

where $(W_t)_{t\geq 0}$ is a standard Brownian motion, $(N_t)_{t\geq 0}$ is a Poisson process with intensity λ, and $(U_j)_{j\geq 1}$ is a sequence of independent, identically distributed random variables, with values in $]-1, +\infty[$ and law $\mu(dx)$. The σ-algebras generated by $(W_t)_{t\geq 0}$, $(N_t)_{t\geq 0}$, $(U_j)_{j\geq 1}$ are supposed to be independent.

To simulate this process at times $n\Delta t$, we notice that

$$X_{n\Delta t} = x \times (X_{\Delta t}/x) \times (X_{2\Delta t}/X_{\Delta t}) \times \cdots \times (X_{n\Delta t}/X_{(n-1)\Delta t}).$$

If we note $Y_k = (X_{k\Delta t}/X_{(k-1)\Delta t})$, we can prove, from the properties of $(N_t)_{t\geq 0}$,

$(W_t)_{t\geq 0}$ and $(U_j)_{j\geq 1}$ that $(Y_k)_{k\geq 1}$ is a sequence of independent random variables with the same law. Since $X_{n\Delta t} = x Y_1 \ldots Y_n$, the simulation of X at times $n\Delta t$ comes down to the simulation of the sequence $(Y_k)_{k\geq 1}$. This sequence being independent and identically distributed, it suffices to know how to simulate $Y_1 = X_{\Delta t}/x$. Then we operate as follows:

- We simulate a standard Gaussian random variable g.

- We simulate a Poisson random variable with parameter $\lambda\Delta t$: N.

- If $N = n$, we simulate n random variables following the law $\mu(dx)$: $U_1, \ldots, U_n$.

All these variables are assumed to be independent. Then, from equation (8.3), it is clear that the law of

$$\left(\prod_{j=1}^{N}(1 + U_j) \right) e^{(\mu-\sigma^2/2)\Delta t+\sigma\sqrt{\Delta t}g}$$

is identical to the law of Y_1.

8.2 Some useful algorithms

In this section, we have gathered some widely used algorithms for the pricing of options.

8.2.1 Approximation of the distribution function of a Gaussian variable

We saw in Chapter 4 that the pricing of many classical options requires the calculation of

$$N(x) = \mathbf{P}(X \leq x) = \int_{-\infty}^{x} e^{-\frac{x^2}{2}} \frac{dx}{\sqrt{2\pi}}.$$

where X is a standard Gaussian random variable. Due to the importance of this function in the pricing of options, we give two approximation formulae from Abramowitz and Stegun (1970).

The first approximation is accurate to 10^{-7}, but it uses the exponential function. If $x > 0$

$$
\begin{aligned}
p &= 0.231\,641\,900 \\
b_1 &= 0.319\,381\,530 \\
b_2 &= -0.356\,563\,782 \\
b_3 &= 1.781\,477\,937 \\
b_4 &= -1.821\,255\,978 \\
b_5 &= 1.330\,274\,429 \\
t &= 1/(1 + px)
\end{aligned}
$$

$$N(x) \approx 1 - \frac{1}{\sqrt{2\pi}}e^{-\frac{x^2}{2}}(b_1 t + b_2 t^2 + b_3 t^3 + b_4 t^4 + b_5 t^5).$$

The second approximation is accurate to 10^{-3} but it involves only a ratio as opposed to an exponential. If $x > 0$

$$c_1 = 0.196\,854$$
$$c_2 = 0.115\,194$$
$$c_3 = 0.000\,344$$
$$c_4 = 0.019\,527$$

$$N(x) \approx 1 - \frac{1}{2}(1 + c_1 x + c_2 x^2 + c_3 x^3 + c_4 x^4)^{-4}.$$

8.2.2 Implementation of the Brennan and Schwartz method

The following program prices an American put using the method described in Chapter 5, Section 5.3.2: we make a logarithmic change of variable, we discretise the parabolic inequality using a totally implicit method and finally solve the inequality in infinite dimensions using the algorithm described on page 116.

```
CONST
   PriceStepNb  =  200;
   TimeStepNb =  200;
   Accuracy    = 0.01;
   DaysInYearNb = 360;

TYPE
   Date = INTEGER;
   Amount = REAL;
   AmericanPut = RECORD
ContractDate    : Date;(* in days *)
MaturityDate   : Date;(* in days *)
StrikePrice   : Amount;
END;
   vector = ARRAY[1..PriceStepNb] OF REAL;
   Model = RECORD
r       : REAL; (* annual riskless interest rate *)
sigma   : REAL; (* annual volatility *)
x0      : REAL; (* initial value of the SDE *)
END;

FUNCTION PutObstacle(x : REAL;Opt : AmericanPut) : REAL;
VAR u : REAL;
BEGIN
     u := Opt.StrikePrice - exp(x);
     IF u > 0 THEN PutObstacle := u ELSE PutObstacle := 0.0;
END;

FUNCTION Price(t : Date; x : Amount;
 option : AmericanPut; model : Model) : REAL;
(*
   prices the 'option' for the 'model'
   at time 't' if the price of the underlying at this time
   is 'x'.
*)
VAR
   Obst,A,B,C,G : vector;
   alpha,beta,gamma,h,k,vv,temp,r,y,delta,Time,l : REAL;
   Index,PriceIndex,TimeIndex : INTEGER;
BEGIN
   Time := (option.MaturityDate - t) / DaysInYearNb;
   k := Time / TimeStepNb;
   r := model.r;
```

```
        vv := model.sigma * model.sigma;

        l := (model.sigma * sqrt(Temps) * sqrt(ln(1/Accuracy))) + abs(r - vv / 2) *
Time);
        h := 2 * l / PriceStepNb;
        writeln(l:5:3,' ',ln(2):5:3);

        alpha := k * (- vv / (2.0 * h * h) + (r - vv / 2.0) / (2.0 * h));
        beta := 1 + k * (r + vv / (h * h));
        gamma := k * (- vv / (2.0 * h * h) - (r - vv / 2.0) / (2.0 * h));
        FOR PriceIndex:=1 TO PriceStepNb DO BEGIN
A[PriceIndex] :=   alpha;
B[PriceIndex] :=   beta;
C[PriceIndex] :=   gamma;
END;
        B[1]                := beta + alpha;
        B[PriceStepNb]   := beta + gamma;
        G[PriceIndex]    := 0.0;

        B[PriceStepNb] := B[PriceStepNb];
        FOR PriceIndex:=PriceStepNb-1 DOWNTO 1 DO
B[PriceIndex] := B[PriceIndex] - C[PriceIndex] * A[PriceIndex+1] /
            B[PriceIndex+1];
        FOR PriceIndex:= 1 TO PriceStepNb DO A[PriceIndex] := A[PriceIndex] /
            B[PriceIndex];
        FOR PriceIndex:= 1 TO PriceStepNb - 1 DO C[PriceIndex] := C[PriceIndex] /
            B[PriceIndex+1];

        y := ln(x);
        FOR PriceIndex:=1 TO PriceStepNb DO Obst[PriceIndex] := PutObstacle(y - l +
PriceIndex * h , option );
        FOR PriceIndex:=1 TO PriceStepNb DO G[PriceIndex] := Obst[PriceIndex];

        FOR TimeIndex:=1 TO TimeStepNb DO BEGIN
FOR PriceIndex := PriceStepNb-1 DOWNTO 1 DO
        G[PriceIndex] := G[PriceIndex] - C[PriceIndex] * G[PriceIndex+1];


G[1] := G[1] / B[1];
FOR PriceIndex:=2 TO PriceStepNb DO BEGIN
 G[PriceIndex] := G[PriceIndex] / B[PriceIndex] - A[PriceIndex]* G[Price
Index-1];
 temp := Obst[PriceIndex];
 IF G[PriceIndex] < temp THEN G[PriceIndex] := temp;
END;
        END;
        Index := PriceStepNb DIV 2;
        delta   := (G[Indice+1] - G[Index]) / h;
        Prix := G[Index]+ delta*(Index * h - l);
END;
```

8.3 Exercises

Exercise 44 Let X and Y be two standard Gaussian random variables; derive the joint law of $(\sqrt{X^2 + Y^2}, \text{arctg}(Y/X))$. Deduce that, if U_1 and U_2 are two independent uniform random variables on $[0, 1]$, the random variables $\sqrt{-2\log(U_1)}\cos(2\pi U_2)$ and $\sqrt{-2\log(U_1)}\sin(2\pi U_2)$ are independent and follow a standard Gaussian law.

Exercise 45 Let f be a function from $\mathbb{R}$ to $\mathbb{R}$, such that $f(x) > 0$ for all x, and such that $\int_{-\infty}^{+\infty} f(x)dx = 1$. We want to simulate a random variable X with law

$f(x)dx$. We set $F(u) = \int_{-\infty}^{u} f(x)dx$. Prove that if U is a uniform random variable on $[0, 1]$, then the law of $F^{-1}(U)$ is $f(x)dx$. Deduce a method of simulation of X.

Exercise 46 We model a risky asset S_t by the stochastic differential equation

$$\begin{cases} dS_t &= S_t(\mu dt + \sigma dW_t) \\ S_0 &= x, \end{cases}$$

where $(W_t)_{t\geq 0}$ is a standard Brownian motion, σ the volatility and r is the riskless interest rate. Propose a method of simulation to approximate

$$E\left(e^{-rT}\left(\frac{1}{T}\int_0^T S_s ds - S_T\right)_+\right).$$

Give an interpretation for the final value in terms of option.

Exercise 47 The aim of this exercise is to study the influence of the hedging frequency on the variance of a portfolio of options. The underlying asset of the options is described by the Black-Scholes model

$$\begin{cases} dS_t &= S_t(\mu dt + \sigma dW_t) \\ S_0 &= x, \end{cases}$$

$(W_t)_{t\geq 0}$ represents a standard Brownian motion, σ the annual volatility and r the riskless interest rate. Further on we will fix $r = 10\%/\text{year}$, $\sigma = 20\%/\sqrt{\text{year}} = 0.2$ and $x = 100$.

Being 'delta neutral' means that we compensate the total delta of the portfolio by trading the adequate amount of underlying asset.

In the following, the options have 3 months to maturity and are contingent on one unit of asset. We will choose one of the following combinations of options:

- **Bull spread**: long a call with strike price 90 (written as 90 call) and short a 110 call with same maturity.
- **Strangle**: short a 90 put and short a 110 call.
- **Condor**: short a 90 call, long a 95 call and a 105 call and finally short a 110 call.
- **Put ratio backspread**: short a 110 put and long 3 90 puts.

First we suppose that $\mu = r$. Write a program which:

- Simulates the asset described previously.
- Calculates the mean and variance of the discounted final value of the portfolio in the following cases:
  - We do not hedge: we sell the option, get the premium, we wait for three months, we take into account the exercise of the option sold and we evaluate the portfolio.
  - We hedge immediately after selling the option, then we do nothing.

- We hedge immediately after selling the option, then every month.
- We hedge immediately after selling the option, then every 10 days.
- We hedge immediately after selling the option, then every day.

Investigate the influence of the discretisation frequency.

Now consider the previous simulation assuming that $\mu \neq r$ (take values of μ bigger and smaller than r). Are there arbitrage opportunities?

Exercise 48 We suppose that $(W_t)_{t\geq0}$ is a standard Brownian motion and that $(U_i)_{i\geq1}$ is a sequence of independent random variables taking values $+1$ or -1 with probability $1/2$. We set $S_n = X_1 + \cdots + X_n$.

1. Prove that, if $X_t^n = S_{[nt]}/\sqrt{n}$, X_t^n converges in law to W_t.

2. Let t and s be non-negative; using the fact that the random variable $X_{t+s}^n - X_t^n$ is independent of X_t^n, prove that the pair (X_{t+s}^n, X_t^n) converges in law to (W_{t+s}, W_t).

3. If $0 < t_1 < \cdots < t_p$, show that $(X_{t_1}^n, \ldots, X_{t_p}^n)$ converges in law to $(W_{t_1}, \ldots, W_{t_p})$.

Appendix

A.1 Normal random variables

In this section, we recall the main properties of Gaussian variables. The following results are proved in Bouleau (1986), Chapter VI, Section 9.

A.1.1 Scalar normal variables

A real random variable X is a standard normal variable if its probability density function is equal to

$$n(x) = \frac{1}{\sqrt{2\pi}} \exp\left(-\frac{x^2}{2}\right).$$

If X is a standard normal variable and m and σ are two real numbers, then the variable $Y = m + \sigma X$ is normal with mean m and variance σ^2. Its law is denoted by $\mathcal{N}(m, \sigma^2)$ (it does not depend on the sign of σ since X and $-X$ have the same law). If $\sigma \neq 0$, the density of Y is

$$\frac{1}{\sqrt{2\pi\sigma^2}} \exp\left(-\frac{(x - m)^2}{2\sigma^2}\right).$$

If $\sigma = 0$, the law of Y is the Dirac measure in m and therefore it does not have a density. It is sometimes called 'degenerate normal variable'.

If X is a standard normal variable, we can prove that for any complex number z, we have

$$\mathbf{E}\left(e^{zX}\right) = e^{\frac{z^2}{2}}.$$

Thus, the characteristic function of X is given by $\phi_X(u) = e^{-u^2/2}$ and for Y, $\phi_Y(u) = e^{ium}e^{-u^2\sigma^2/2}$. It is sometimes useful to know that if X is a standard normal variable, we have $\mathbf{P}(|X| > 1,96\ldots) = 0,05$ and $\mathbf{P}(|X| > 2,6\ldots) = 0,01$. For large values of $t > 0$, the following approximation is handy:

$$\mathbf{P}(X > t) = \frac{1}{\sqrt{2\pi}} \int_t^\infty e^{-x^2/2}dx \leq \frac{1}{t\sqrt{2\pi}} \int_t^\infty xe^{-x^2/2}dx = \frac{e^{-t^2/2}}{t\sqrt{2\pi}}.$$

Finally, one should know that there exist very good approximations of the cumulative normal distribution (cf. Chapter 8) as well as statistical tables.

A.1.2 Multivariate normal variables

Definition A.1.1 *A random variable* $X = (X_1, \ldots, X_d)$ *in* $\mathbb{R}^d$ *is a Gaussian vector if for any sequence of real numbers* $a_1, \ldots, a_d$, *the scalar random variable* $\sum_{i=1}^{d} a_i X_i$ *is normal.*

The components $X_1, \ldots, X_d$ of a Gaussian vector are obviously normal, but the fact that each component of a vector is a normal random variable does not imply that the vector is normal. However, if $X_1, X_2, \ldots, X_d$ are real-valued, normal, *independent* random variables, then the vector $(X_1, \ldots, X_d)$ is normal.

The covariance matrix of a random vector $X = (X_1, \ldots, X_d)$ is the matrix $\Gamma(X) = (\sigma_{ij})_{1 \leq i,j \leq d}$ whose coefficients are equal to

$$\sigma_{ij} = \text{cov}(X_i, X_j) = \mathbf{E}\left[(X_i - \mathbf{E}(X_i))(X_j - \mathbf{E}(X_j))\right].$$

It is well known that if the random variables $X_1, \ldots, X_d$ are independent, the matrix $\Gamma(X)$ is diagonal, but the converse is generally wrong, except in the Gaussian case:

Theorem A.1.2 *Let* $X = (X_1, \ldots, X_d)$ *be a Gaussian vector in* $\mathbb{R}^d$. *The random variables* $X_1, \ldots, X_d$ *are independent if and only if the covariance matrix* X *is diagonal.*

The reader should consult Bouleau (1986), Chapter VI, p. 155, for a proof of this result.

Remark A.1.3 The importance of normal random variables in modelling comes partly from the central limit theorem (cf. Bouleau (1986), Chapter VII, Section 4). The reader ought to refer to Dacunha-Castelle and Duflo (1986) (Chapter 5) for problems of estimation and to Chapter 8 for problems of simulation.

A.2 Conditional expectation

A.2.1 Examples of σ-algebras

Let us consider a measurable space $(\Omega, \mathcal{A})$ and a partition $B_1, B_2, \ldots, B_n$, with n events in $\mathcal{A}$. The set $\mathcal{B}$ containing the elements of $\mathcal{A}$ which are either empty or that can be written as $B_{i_1} \cup B_{i_2} \cup \cdots \cup B_{i_k}$, where $i_1, \ldots, i_k \in \{1, \ldots, n\}$, is a finite sub-$\sigma$-algebra of $\mathcal{A}$. It is the σ-algebra generated by the sequence of B_i.

Conversely, to any finite sub-σ-algebra $\mathcal{B}$ of $\mathcal{A}$, we can associate a finite partition $(B_1, \ldots, B_n)$ of Ω where $\mathcal{B}$ is generated by the elements B_i of $\mathcal{A}$: B_i are the non-empty elements of $\mathcal{B}$ which contain only themselves and the empty set. They are called atoms of $\mathcal{B}$. There is a one-to-one mapping from the set of finite sub-σ-algebras of $\mathcal{A}$ onto the set of partitions of Ω by elements of $\mathcal{A}$. One should notice that if $\mathcal{B}$ is a sub-σ-algebra of $\mathcal{A}$, a map from Ω to $\mathbb{R}$ (and its Borel σ-algebra) is $\mathcal{B}$-measurable if and only if it is constant on each atom of $\mathcal{B}$.

Let us now consider a random variable X defined on $(\Omega, \mathcal{A})$ with values in a measurable space $(E, \mathcal{E})$. The σ-algebra generated by X is the smallest σ-algebra for which X is measurable: it is denoted by $\sigma(X)$. It is obviously included in $\mathcal{A}$ and it is easy to show that

$$\sigma(X) = \left\{ A \in \mathcal{A} | \exists B \in \mathcal{E}, A = X^{-1}(B) = \{X \in B\} \right\}.$$

We can prove that a random variable Y from $(\Omega, \mathcal{A})$ to $(F, \mathcal{F})$ is $\sigma(X)$-measurable if and only if it can be written as

$$Y = f \circ X,$$

where f is a measurable map from $(E, \mathcal{E})$ to $(F, \mathcal{F})$ (cf. Bouleau (1986), p. 101-102). In other words, $\sigma(X)$-measurable random variables are the measurable functions of X.

A.2.2 Properties of the conditional expectation

Let $(\Omega, \mathcal{A}, \mathbf{P})$ be a probability space and $\mathcal{B}$ a σ-algebra included in $\mathcal{A}$. The definition of the conditional expectation is based on the following theorem (refer to Bouleau (1986), Chapter 8):

Theorem A.2.1 *For any real integrable random variable X, there exists a real integrable $\mathcal{B}$-measurable random variable Y such that*

$$\forall B \in \mathcal{B} \quad \mathbf{E}(X\mathbf{1}_B) = \mathbf{E}(Y\mathbf{1}_B).$$

If $\tilde{Y}$ is another random variable with these properties then $\tilde{Y} = Y$ $\mathbf{P}$ a.s. Y is the conditional expectation of X given $\mathcal{B}$ and it is denoted by $\mathbf{E}(X|\mathcal{B})$.

If $\mathcal{B}$ is a finite sub-σ-algebra, with atoms $B_1, \ldots, B_n$,

$$\mathbf{E}(X|\mathcal{B}) = \sum_i \mathbf{E}(X\mathbf{1}_{B_i})/\mathbf{P}(B_i)\mathbf{1}_{B_i},$$

where we sum on the atoms with strictly positive probability. Consequently, on each atom B_i, $\mathbf{E}(X|\mathcal{B})$ is the mean value of X on B_i. As far as the trivial σ-algebra is concerned $(\mathcal{B} = \{\emptyset, \Omega\})$, we have $\mathbf{E}(X|\mathcal{B}) = \mathbf{E}(X)$.

The computations involving conditional expectations are based on the following properties:

1. If X is $\mathcal{B}$-measurable, $\mathbf{E}(X|\mathcal{B}) = X$, a.s.

2. $\mathbf{E}(\mathbf{E}(X|\mathcal{B})) = \mathbf{E}(X)$.

3. For any bounded, $\mathcal{B}$-measurable random variable Z, $\mathbf{E}(Z\mathbf{E}(X|\mathcal{B})) = \mathbf{E}(ZX)$.

4. Linearity:

$$\mathbf{E}(\lambda X + \mu Y|\mathcal{B}) = \lambda \mathbf{E}(X|\mathcal{B}) + \mu \mathbf{E}(Y|\mathcal{B}) \text{ a.s.}$$

5. Positivity: if $X \geq 0$, then $\mathbf{E}(X|\mathcal{B}) \geq 0$ a.s. and more generally, $X \geq Y \Rightarrow \mathbf{E}(X|\mathcal{B}) \geq \mathbf{E}(Y|\mathcal{B})$ a.s. It follows from this property that

$$|\mathbf{E}(X|\mathcal{B})| \leq \mathbf{E}(|X||\mathcal{B}) \text{ a.s.}$$

and therefore $\|\mathbf{E}(X|\mathcal{B})\|_{L^1(\Omega)} \leq \|X\|_{L^1(\Omega)}$.

6. If $\mathcal{C}$ is a sub-σ-algebra of $\mathcal{B}$, then

$$\mathbf{E}\left(\mathbf{E}\left(X|\mathcal{B}\right)|\mathcal{C}\right) = \mathbf{E}\left(X|\mathcal{C}\right) \text{ a.s.}$$

7. If Z is $\mathcal{B}$-measurable and bounded, $\mathbf{E}\left(ZX|\mathcal{B}\right) = Z\mathbf{E}\left(X|\mathcal{B}\right)$ a.s.

8. If X is independent of $\mathcal{B}$ then $\mathbf{E}\left(X|\mathcal{B}\right) = \mathbf{E}\left(X\right)$ a.s.

The converse property is not true but we have the following result.

Proposition A.2.2 *Let X be a real random variable. X is independent of the σ-algebra $\mathcal{B}$ if and only if*

$$\forall u \in \mathbb{R} \quad \mathbf{E}\left(e^{iuX}|\mathcal{B}\right) = \mathbf{E}\left(e^{iuX}\right) \text{ a.s.} \tag{A.1}$$

Proof. Given the Property 8. above, we just need to prove that (A.1) implies that X is independent of $\mathcal{B}$.

If $\mathbf{E}\left(e^{iuX}|\mathcal{B}\right) = \mathbf{E}\left(e^{iuX}\right)$ then, by definition of the conditional expectation, for all $B \in \mathcal{B}$, $\mathbf{E}\left(e^{iuX}\mathbf{1}_B\right) = \mathbf{E}\left(e^{iuX}\right)\mathbf{P}(B)$. If $\mathbf{P}(B) \neq 0$, we can write

$$\mathbf{E}\left(e^{iuX}\frac{\mathbf{1}_B}{\mathbf{P}(B)}\right) = \mathbf{E}\left(e^{iuX}\right).$$

This equality means that the characteristic function of X is identical under measure $\mathbf{P}$ and measure $\mathbf{Q}$ where the density of $\mathbf{Q}$ with respect to $\mathbf{P}$ is equal to $\mathbf{1}_B/\mathbf{P}(B)$. The equality of characteristic functions implies the equality of probability laws and consequently

$$\mathbf{E}\left(f(X)\frac{\mathbf{1}_B}{\mathbf{P}(B)}\right) = \mathbf{E}\left(f(X)\right),$$

for any bounded Borel function f, hence the independence. $\qquad\square$

Remark A.2.3 If X is square integrable, so is $\mathbf{E}(X|\mathcal{B})$, and $\mathbf{E}(X|\mathcal{B})$ coincides with the orthogonal projection of X on $\mathbf{L}^2(\Omega, \mathcal{B}, \mathbf{P})$, which is a closed subspace of $\mathbf{L}^2(\Omega, \mathcal{A}, \mathbf{P})$, together with the scalar product $(X, Y) \mapsto \mathbf{E}(XY)$ (cf. Bouleau (1986), Chapter VIII, Section 2). The conditional expectation of X given $\mathcal{B}$ is the least-square-best $\mathcal{B}$-measurable predictor of X. In particular, if $\mathcal{B}$ is the σ-algebra generated by a random variable ξ, the conditional expectation $\mathbf{E}(X|\mathcal{B})$ is noted $\mathbf{E}(X|\xi)$, and it is the best approximation of X by a function of ξ, since $\sigma(\xi)$-measurable random variables are the measurable functions of ξ. Notice that by Pythagoras' theorem, we know that $\|\mathbf{E}(X|\mathcal{B})\|_{L^2(\Omega)} \leq \|X\|_{L^2(\Omega)}$.

Remark A.2.4 We can define $\mathbf{E}(X|\mathcal{B})$ for any non-negative random variable X (without integrability condition). Then $\mathbf{E}(XZ) = \mathbf{E}\left(\mathbf{E}(X|\mathcal{B})Z\right)$, for any $\mathcal{B}$-measurable non-negative random variable Z. The rules are basically the same as in the integrable case (see Dacunha-Castelle and Duflo (1982), Chapter 6).

A.2.3 *Computations of conditional expectations*

The following proposition is crucial and is used quite often in this book.

Proposition A.2.5 *Let us consider a $\mathcal{B}$-measurable random variable X taking values in $(E, \mathcal{E})$ and Y, a random variable independent of $\mathcal{B}$ with values in $(F, \mathcal{F})$. For any Borel function Φ non-negative (or bounded) on $(E \times F, \mathcal{E} \otimes \mathcal{F})$, the function φ defined by*

$$\forall x \in E \quad \varphi(x) = \mathbf{E}\left(\Phi(x, Y)\right)$$

is a Borel function on $(E, \mathcal{E})$ and we have

$$\mathbf{E}\left(\Phi(X, Y)|\mathcal{B}\right) = \varphi(X) \ a.s.$$

In other words, under the previous assumptions, we can compute $\mathbf{E}\left(\Phi(X, Y)|\mathcal{B}\right)$ as if X was a constant.

Proof. Let us denote by $\mathbf{P}_Y$ the law of Y. We have

$$\varphi(x) = \int_F \Phi(x, y) d\mathbf{P}_Y(y)$$

and the measurability of φ is a consequence of the Fubini theorem. Let Z be a non-negative $\mathcal{B}$-measurable random variable (for example $Z = \mathbf{1}_B$, with $B \in \mathcal{B}$). If we denote by $\mathbf{P}_{X,Z}$ the law of (X, Z), it follows from the independence between Y and (X, Z) that,

$$
\begin{aligned}
\mathbf{E}\left(\Phi(X, Y)Z\right) &= \int \int \Phi(x, y) z \, d\mathbf{P}_{X,Z}(x, z) d\mathbf{P}_Y(y) \\
&= \int \left(\int \Phi(x, y) d\mathbf{P}_Y(y) \right) z \, d\mathbf{P}_{X,Z}(x, z) \\
&= \int \varphi(x) z \, d\mathbf{P}_{X,Z}(x, z) \\
&= \mathbf{E}\left(\varphi(X)Z\right),
\end{aligned}
$$

which completes the proof. $\square$

Remark A.2.6 In the Gaussian case, the computation of a conditional expectation is particularly simple. Indeed, if $(Y, X_1, X_2, \ldots, X_n)$ is a normal vector (in $\mathbb{R}^{n+1}$), the conditional expectation $Z = \mathbf{E}\left(Y|X_1, \ldots, X_n\right)$ has the following form

$$Z = c_0 + \sum_{i=1}^{n} c_i X_i,$$

where c_i are real constant numbers. This means that the function of X_i which approximates Y in the least-square sense is linear. On top of that, we can compute Z by projecting the random variable Y in L^2 on the linear subspace generated by $\mathbf{1}$ and the X_i's (cf. Bouleau (1986), Chapter 8, Section 5).

A.3 Separation of convex sets

In this section, we state the theorem of separation of convex sets that we use in the first chapter. For more details, the diligent reader can refer to Dudley (1989) p. 152 or Minoux (1983).

Theorem A.3.1 *Let C be a closed convex set which does not contain the origin. Then there exists a real linear functional ξ defined on $\mathbb{R}^n$ and $\alpha > 0$ such that*

$$\forall x \in C \quad \xi(x) \geq \alpha.$$

In particular, the hyperplane $\xi(x) = 0$ does not intersect C.

Proof. Let λ be a non-negative real number such that the closed ball $B(\lambda)$ with centre at the origin and radius λ intersects C. Let x_0 be the point where the map $x \mapsto \|x\|$ achieves its minimum (where $\|\cdot\|$ is the Euclidean norm) on the compact set $C \cap B(\lambda)$. It follows immediately that

$$\forall x \in C \quad \|x\| \geq \|x_0\|.$$

The vector x_0 is nothing but the projection of the origin on the closed convex set C. If we consider $x \in C$, then for all $t \in [0, 1]$, $x_0 + t(x - x_0) \in C$, since C is convex. By expanding the following inequality

$$\|x_0 + t(x - x_0)\|^2 \geq \|x_0\|^2,$$

it yields $x_0.x \geq \|x_0\|^2 > 0$ for any $x \in C$, where $x_0.x$ denotes the scalar product of x_0 and x. This completes the proof.   □

Theorem A.3.2 *Let us consider a compact convex set K and a vector subspace V of $\mathbb{R}^n$. If V and K are disjoint, there exists a linear functional ξ defined on $\mathbb{R}^n$, satisfying the following conditions:*

1. $\forall x \in K \quad \xi(x) > 0$.

2. $\forall x \in V \quad \xi(x) = 0$.

Therefore, the subspace V is included in a hyperplane that does not intersect K.

Proof. The set

$$C = K - V = \{x \in \mathbb{R}^n \mid \exists(y, z) \in K \times V, x = y - z\}$$

is convex, closed (because V is closed and K is compact) and does not contain the origin. By Theorem A.3.1, we can find a linear functional ξ defined on $\mathbb{R}^n$ and a certain $\alpha > 0$ such that

$$\forall x \in C \quad \xi(x) \geq \alpha.$$

Hence

$$\forall y \in K, \ \forall z \in V, \ \xi(y) - \xi(z) \geq \alpha. \tag{A.2}$$

For a fixed y, we can apply (A.2) to λz, with $\lambda \in \mathbb{R}$ to obtain: $\forall z \in V, \ \xi(z) = 0$, thus $\forall y \in K, \ \xi(y) \geq \alpha$,   □

References

Abramowitz, M. and I.A. Stegun (eds), *Handbook of Mathematical Functions*, 9th printing, 1970.

Artzner, P. and F. Delbaen, Term structure of interest rates: The martingale approach, *Advances in Applied Mathematics* 10 (1989), pp. 95-129.

Bachelier, L., Théorie de la spéculation, *Ann. Sci. Ecole Norm. Sup.*, 17 (1900), pp. 21-86.

Barone-Adesi, G. and R. Whaley, Efficient analytic approximation of American option values, *J. of Finance*, 42 (1987), pp. 301-320.

Bensoussan, A., On the theory of option pricing, *Acta Appl. Math.*, 2 (1984) pp. 139-158.

Bensoussan, A. and J.L. Lions, *Applications des inéquations variationnelles en contrôle stochastique*, Dunod, 1978.

Bensoussan, A. and J.L. Lions *Applications of Variational Inequalities in Stochastic Control*, North-Holland, 1982.

Black, F. and M. Scholes, The pricing of options and corporate liabilities, *Journal of Political Economy*, 81 (1973), pp. 635-654.

Brennan, M.J. and E.S. Schwartz, The valuation of the American put option, *J. of Finance*, 32, 1977, pp. 449-462.

Brennan, M.J. and E.S. Schwartz, A continuous time approach to the pricing of bonds, *J. of Banking and Finance*, 3 (1979), pp. 133-155.

Bouleau, N., *Probabilités de l'Ingénieur*, Hermann, 1986.

Bouleau, N., *Processus Stochastiques et Applications*, Hermann, 1988.

Bouleau, N. and D.Lamberton, Residual risks and hedging strategies in Markovian markets, *Stoch. Proc. and Appl.*, 33 (1989), pp. 131-150.

CERMA, Sur les risques résiduels des stratégies de couverture d'actifs conditionnels, *Comptes Rendus de l'Académie des Sciences*, 307 (1988), pp. 625-630.

Chateau, O., *Quelques remarques sur les processus à accroissements indépendants et stationnaires, et la subordination au sens de Bochner*, Thèse de l'Université Paris VI, 1990.

Ciarlet, P.G., *Une Introduction à l'analyse numérique matricielle et à l'optimisation*, Masson, 1988.

Ciarlet, P.G. and J.L. Lions, *Handbook of Numerical Analysis*, Volume I, Part 1 (1990), North Holland.

Courtadon, G., The pricing of options on default-free bonds, *J. of Financial and Quant. Anal.*, 17 (1982), pp. 301-329.

Cox, J.C., J.E. Ingersoll and S.A. Ross, A theory of the term structure of interest rates, *Econometrica*, 53 (1985), pp. 385-407.

Cox, J.C. and M. Rubinstein, *Options Markets*, Prentice–Hall, 1985.

Dalang, R.C., A. Morton and W. Willinger, Equivalent martingale measures and no-arbitrage in stochastic securities market models, *Stochastics and Stochastics Reports*, vol. 29, (2) (1990), pp. 185-202.

Dacunha-Castelle, D. and M. Duflo, *Probability and Statistics*, volume 1, Springer-Verlag, 1986.

Dacunha-Castelle, D. and M. Duflo, *Probability and Statistics*, volume 2, Springer-Verlag, 1986.

Delbaen, F. and W. Schachermayer, A general version of the fundamental theorem of asset pricing, *Math. Ann. 300* (1994), pp. 463-520.

Dudley, R.M., *Real Analysis and Probability*, Wadsworth & Brooks/Cole, 1989.

Duffie, D., *Security Markets, Stochastic Models*, Academic Press, 1988.

El Karoui, N., *Les aspects probabilistes du contrôle stochastique*, Lecture Notes in Mathematics 876, pp. 72-238, Springer-Verlag, 1981.

El Karoui, N. and J.C. Rochet, A pricing formula for options on coupon-bonds, *Cahier de recherche du GREMAQ-CRES*, no. 8925, 1989.

El Karoui, N. and M.C. Quenez, Dynamic programming and pricing of contingent claims in an incomplete market, *S.I.A.M. J. Control and Optimization* 33 (1995), pp. 29-66.

Föllmer, H. and D. Sondermann, Hedging of non-redundant contingent claims, *Contributions to Mathematical Economics in Honor of Gerard Debreu*, W.Hildebrand and A. Mas-Colell (eds), North-Holland, Amsterdam, 1986.

Föllmer, H. and D. Sondermann, Hedging of contingent claims under incomplete information, *Applied Stochastic Analysis*, M.H.A. Davis and R.J. Elliott (eds), Gordon & Breach, 1990.

Föllmer, H. and M. Schweizer, "Hedging of contingent claims under incomplete information", *Applied Stochastic Analysis*, M.H.A. Davis and R.J. Elliott (eds), Stochastics Monographs, vol. 5, Gordon & Breach (1991), pp. 389-414.

Friedman, A., *Stochastic Differential Equations and Applications*, Academic Press, 1975.

Gard, T., *Introduction to Stochastic Differential Equations*, Marcel Dekker, 1988.

Garman, M.B. and S.W. Kohlhagen, Foreign currency option values, *J. of International Money and Finance*, 2 (1983), pp. 231-237.

Gihman, I.I. and A.V. Skorohod, *Introduction à la Théorie des Processus Aléatoires*, Mir, 1980.

Glowinsky, R., J.L. Lions and R. Trémolières, *Analyse numérique des inéquations variationnelles*, Dunod, 1976.

Harrison, M.J. and D.M. Kreps, Martingales and arbitrage in multiperiod securities markets, *J. of Economic Theory*, 29 (1979), pp. 381-408.

Harrison, M.J. and S.R. Pliska, Martingales and stochastic integrals in the theory of continuous trading, *Stochastic Processes and their Applications*, 11 (1981), pp. 215-260.

Harrison, M.J. and S.R.Pliska, A stochastic calculus model of continuous trading: complete markets, *Stochastic Processes and their Applications*, 15 (1983), pp. 313-316.

Heath, D., A. Jarrow and A. Morton, Bond pricing and the term structure of interest rates, *preprint*, 1987.

Ho, T.S. and S.B. Lee, Term structure movements and pricing interest rate contingent claims, *J. of Finance*, 41 (1986), pp. 1011-1029.

Huang, C.F. and R.H. Litzenberger, *Foundations for Financial Economics*, North-Holland, 1988.

Ikeda, N. and S. Watanabe, *Stochastic Differential Equations and Diffusion Processes*, North-Holland, 1981.

Jaillet, P., D. Lamberton and B. Lapeyre, Variational inequalities and the pricing of American options, *Acta Applicandae Mathematicae*, 21 (1990), pp. 263-289.

Jamshidian, F., An exact bond pricing formula, *Journal of Finance*, 44 (1989), pp. 205-209.

Karatzas, I., On the pricing of American options, *Applied Mathematics and Optimization*, 17 (1988),pp. 37-60.

Karatzas, I. and S.E. Shreve, *Brownian Motion and Stochastic Calculus*, Springer-Verlag, 1988.

Kloeden, P.E. and E. Platen, *Numerical Solution of Stochastic Differential Equations*, Springer-Verlag, 1992.

Kushner, H.J., *Probability Methods for Approximations in Stochastic Control and for Elliptic Equations*, Academic Press, 1977.

Knuth, D.E., *The Art of Computer Programming*, Vol. 2: Seminumerical algorithms, Addison-Wesley, 1981.

Lamberton, D. and G. Pagès, Sur l'approximation des réduites, *Annales de l'IHP*, 26 (1990), pp. 331-355.

L'Ecuyer, P., Random numbers for simulation, Communications of the ACM, October 1990, vol. 33 (10).

MacMillan, L., Analytic approximation for the American put price, *Advances in Futures and Options Research*, 1 (1986), pp. 119-139.

Merton, R.C., Theory of rational option pricing, *Bell J. of Econ. and Management Sci.*, 4 (1973), pp. 141-183.

Merton, R.C., Option pricing when underlying stock returns are discontinuous, *J. of Financial Economics*, 3 (1976), pp. 125-144.

Minoux, M., *Programmation mathématique*, 2 vols, Dunod, 1983.

Morton, A.J., *Arbitrage and Martingales*, Ph.D. thesis, Cornell University, 1989.

Neveu, J., *Martingales à temps discret*, Masson, 1972.

Pardoux, E. and D. Talay, Discretization and simulation of stochastic differential equations, *Acta Applicandae Mathematicae*, 3 (1985), pp. 23-47.

Raviart, P.A. and J.M. Thomas, *Introduction à l'analyse numérique des équations aux dérivées partielles*, Masson, 1983.

Rogers, L.C.G. and D. Williams, *Diffusions, Markov Processes and Martingales*, Vol. 2, Itô Calculus, John Wiley & Sons, 1987.

Revuz, A. and M. Yor, *Continuous Martingale Calculus*, Springer-Verlag, 1990.

Rubinstein, R.Y., *Simulation and the Monte Carlo Method*, (1981), John Wiley and sons.

Schaefer, S. and E.S. Schwartz, A two-factor model of the term structure: an approximate analytical solution, *J. of Finance and Quant. Anal.*, 19 (1984), pp. 413-424.

Schweizer, M., Option hedging for semi-martingales, *Stoch. Proc. and Appl.*, 37 (1991), pp. 339-363.

Schweizer, M., Mean-variance hedging for general claims, *Annals of Applied Probability*, 2 (1992), pp. 171-179.

Schweizer, M., Approximating random variables by stochastic integrals, *Annals of Probability*, 22 (1993), pp. 1536-1575.

Schweizer, M., On the minimal martingale measure and the Föllmer-Schweizer decomposition, *Stochastic Analysis and Applications*, 13 (1994), pp. 573-599.

Sedgewick, R., *Algorithms*, Addison-Wesley, 1987.

Shiryayev, A.N., *Optimal Stopping Rules*, Springer-Verlag, 1978.

Stricker, C., Arbitrage et lois de martingales, Ann. Inst. Henri Poincarré, 26 (1990), pp. 451-460.

Williams, D., *Probability with Martingales*, Cambridge University Press, 1991.

Zhang, X.L., Numerical analysis of American option pricing in a jump diffusion model, CERMICS, Soc.Gen , 1994.

Index